Basic Coastal Engineering

Ocean Engineering: A Wiley Series

EDITOR:
MICHAEL E. McCORMICK, Ph. D.
U.S. Naval Academy
ASSOCIATE EDITOR:
RAMESWAR BHATTACHARYYA, D. Ingr.
U.S. Naval Academy

Michael E. McCormick	Ocean Engineering Wave Mechanics
John B. Woodward	Marine Gas Turbines
H. O. Berteaux	Buoy Engineering
Clarence S. Clay and Herman Medwin	Acoustical Oceanography: Principles and Applications
F. W. Wheaton	Aquacultural Engineering
Robert M. Sorensen	Basic Coastal Engineering

BASIC COASTAL ENGINEERING

ROBERT M. SORENSEN

Chief, Coastal Structures Branch
U.S. Army Coastal Engineering Research Center
Fort Belvoir, Virginia

and

Adjunct Professor of Engineering
George Washington University
Washington, D.C.

A WILEY-INTERSCIENCE PUBLICATION

JOHN WILEY & SONS, New York · Chichester · Brisbane · Toronto

Library of Congress Cataloging in Publication Data:

Sorenson, Robert M., 1938–
  Basic coastal engineering.

  Ocean engineering, a Wiley series)
  "A Wiley-Interscience publication."
  Includes bibliographical references.
  1. Ocean engineering. 2. Ocean waves.
3. Hydraulic structures. I. Title.

TC1645.S67   627   77-29256
ISBN 0-471-81370-2

Printed in the United States of America

10 9 8 7 6 5 4 3 2

To

Diane, Jonathan, and Jennifer

SERIES PREFACE

Ocean engineering is both old and new. It is old in that man has concerned himself with specific problems in the ocean for thousands of years. Ship building, prevention of beach erosion, and construction of offshore structures are just a few of the specialties that have been developed by engineers over the ages. Until recently, however, these efforts tended to be restricted to specific areas. Within the past decade an attempt has been made to coordinate the activities of all technologists in ocean work, calling the entire field "ocean engineering." Here we have its newness.

Ocean Engineering: A Wiley Series has been created to introduce engineers and scientists to the various areas of ocean engineering. Books in this series are so written as to enable engineers and scientists easily to learn the fundamental principles and techniques of a specialty other than their own. The books can also serve as textbooks in advanced undergraduate and introductory graduate courses. The topics to be covered in this series include ocean engineering wave mechanics, marine corrosion, coastal engineering, dynamics of marine vehicles, offshore structures, and geotechnical or seafloor engineering. We think that this series fills a great need in the literature of ocean technology.

MICHAEL E. MCCORMICK, EDITOR

RAMESWAR BHATTACHARYYA, ASSOCIATE EDITOR

November 1972

PREFACE

This is an introductory text on wave mechanics and other coastal processes fundamental to coastal engineering. Its aim is to provide a background from which the reader can pursue more advanced study of the various theoretical and applied aspects of coastal hydromechanics and coastal engineering design.

This book was written for use in a senior- or postgraduate-level first course in coastal engineering. It should also be suitable for self study by persons having a basic engineering or physical science background. The level of coverage does not require a mathematics or fluid mechanics background beyond that now offered in typical undergraduate civil or mechanical engineering curriculums.

Material presented herein is based on the author's lecture notes from a one semester course taught at the Virginia Polytechnic Institute, Texas A&M University, and at George Washington University. This course was typically followed by courses on wave mechanics, coastal sediment processes, estuary hydromechanics, coastal and offshore structures, and marine foundation engineering.

The topics covered include: (1) derivation and application of the two-dimensional linear wave theory; (2) three-dimensional aspects of wave motion including refraction, diffraction, and reflection; (3) long period water level fluctuations, particularly the tides, tsunamis, basin oscillations, and storm surges; (4) wind wave generation, measurement, analysis, and prediction; (5) wave-structure interaction with emphasis on piles, cables, floating structures, seawalls, and rubble mound structures; (6) littoral zone processes and their interaction with coastal structures; and (7) diffusion in coastal waters, particularly as related to marine outfall design.

Each chapter includes problems designed to demonstrate and expand upon the material covered therein. The references for each chapter were selected partially on the basis of availability and general subject coverage, with the expectation that the reader will use the references to increase his or her understanding of coastal engineering and familiarity with coastal engineering information sources.

I am indebted to E. J. Schmeltz and B. L. McCartney for their reviews of this text and to J. W. Johnson and R. L. Wiegel at the University of California, Berkeley, for introducing me to the subject of coastal engineering.

ROBERT M. SORENSEN

Annandale, Virginia
January 1978

CONTENTS

NOTATION

a	significant structure dimension
a_x, a_y	acceleration components
A	surface or cross-section area
A_i	tidal component amplitude
B, B_0	wave orthogonal spacing
c, c_m, c_0	effluent concentration
C, C_0	wave celerity
C_d	wind drag coefficient
C_D	water drag coefficient
C_g	wave group celerity
C_L	coefficient of lift
C_M	coefficient of mass
C_R	wave reflection coefficient
C_T	wave transmission coefficient
d	water depth
	sediment particle diameter
d_b	wave breaking depth
d_c	channel depth
D	structure diameter
D_x, D_y, D_z	component coefficients of eddy diffusion
E, E_p, E_k	total, potential and kinetic wave energy per unit crest length
$\overline{E}$	average wave energy per unit surface area
f	Coriolis parameter, $2\,\omega\sin\phi$
F	wind fetch length
	total drag and inertia force on a structure
	densimetric Froude number
F_B	buoyant force
F_D	drag force on a structure

F_I	inertia force on a structure
F_L	lift force on a structure
F_{min}	minimum fetch distance
F_0	fall time parameter
F_s	total force per unit structure length
g	acceleration of gravity
g'	apparent acceleration of gravity
h	effluent field vertical thickness
H	wave height
H_b	breaking wave height
H_D	wave height at end of decay distance
H_F	wave height at end of fetch
H_i	incident wave height
	height of i th wave
H_j	height of j th wave in spectrum
H_n	average height of highest n percent of waves
H_0	deep water wave height
H_0'	deep water height equivalent to nearshore height if unaffected by refraction and friction
H_r	reflected wave height
H_{rms}	root mean square wave height
H_s	significant wave height
H_t	transmitted wave height
k	wave number, $2\pi/L$
	oscillation mode
	wind stress coefficient
	virtual mass coefficient
k_c	dieaway coefficient
K	combined wind and bottom stress coefficient
K_D	diffraction coefficient
	armor unit stability coefficient
K_R	refraction coefficient
K_S	shoaling coefficient
l	float separation
L, L_0	wave length
	mean scale of eddy size
L_c	inlet channel length
m	beach slope
M_0	wave force moment around the base of a pile
n	ratio of wave group to phase celerity
	Manning's roughness coefficient

N	number of waves
p	pressure
P	wave power per unit crest length
	precipitation rate
P_l	longshore component of wave energy flux per unit shoreline length
q_x, q_y	directional discharges per unit width
q_s	wind sand transport rate per unit width
Q	effluent discharge
Q_s	longshore sediment transport rate
r	runup ratio
R	vertical elevation of wave runup above SWL
	radius to maximum wind velocity in hurricane
	Reynolds number
S	storm surge setup
	Strouhal number
S_a	initial average effluent dilution in horizontal current
S_c	Coriolis setup
$S_{H^2}(T)$	spectral summation of H^2 as a function of T
S_0	surface effluent dilution at plume centerline
S_p	pressure setup
S_r	armor unit specific gravity
S_w	wind stress setup
S_{ww}	wave setup
t	time
t_d	wind duration
t_t	wave travel time
t_T	tsunami travel time
T	wave period
	time interval
$\bar{T}$	average wave period
T_*	wave and force record lengths
T_D	wave period at end of decay distance
T_e	eddy shedding period
T_F	wave period at end of fetch
T_H	Helmholtz resonant period
T_i	tidal component period
	period of ith wave
T_n	resonant oscillation period
T_s	significant wave period
u, v, w	instantaneous component velocities

u_m	maximum horizontal particle velocity in wave
u_*	shear velocity
U	wind velocity
U_g	geostrophic wind speed
U_R	wind velocity at radius of maximum wind velocity
U_x, U_y	component wind velocities
V	horizontal current velocity
$\mathcal{V}$	structure volume
V_f	particle terminal settling velocity
V_F	forward velocity of hurricane
V_l	average long shore current velocity
w	effluent field width
W	channel width
	weight of stable armor unit
Y	depth to outfall jet centerline
α	angle between wave crest and bottom contour
α_b	wave breaker angle
γ_r	specific weight of armor unit
Δ_i	tidal component phase
ε	vertical ordinate of particle displacement in wave
	surface roughness
ζ	horizontal ordinate of particle displacement in wave
η	surface elevation above SWL
θ	incident wave orthogonal direction
	angle between wind direction and line normal to shore
θ_p	peak force phase angle
λ	basin length
μ	coefficient of static friction
ν	kinematic viscosity
ρ, ρ_f	fluid density
ρ_a	air density
ρ_0	receiving water density
σ	standard deviation
	wave angular frequency, $2\pi/T$
τ_b	sea bottom stress
τ_s	air-sea interface wind stress
ϕ	velocity potential
	sediment size descriptor, $-\log_2(d)$
ω	angular velocity of earth rotation

Basic Coastal Engineering

Chapter

One

COASTAL ENGINEERING

Attempts to solve certain coastal zone problems such as beach erosion and the functional and structural design of harbors date back many centuries. Bruun (1972) discusses some of the early coastal erosion and flooding control activities in Holland, England, and Denmark in his review of the history of coastal defence works as they have developed since the tenth century. Inman (1974), in a study of early harbors around the Mediterranean Sea, found that harbors demonstrating a "very superior 'lay' understanding of waves and currents, which led to development of remarkable concepts in working with natural forces" were constructed as early as 1000–2000 B.C.

Coastal works have usually been the concern of civil and military engineers. The term "coastal engineer" seems to have come into general use as a designation for a definable engineering field in 1950, with the meeting of the First Conference on Coastal Engineering at Long Beach, California. In the preface to the Proceedings of that conference M. P. O'Brien wrote, "It (coastal engineering) is not a new or separate branch of engineering and there is no implication intended that a new breed of engineer or society is in the making. Coastal Engineering is primarily a branch of Civil Engineering which leans heavily on the sciences of oceanography, meteorology, fluid mechanics, electronics, structural mechanics, and others." Among the others, one could add geology and geomorphology, mathematics and statistics, computer science, soil mechanics, and material science.

This definition is still essentially correct. However, coastal engineering has grown in the past two decades to where the U.S. Army Corps of Engineers has a Coastal Engineering Research Center; the American Society of Civil

Engineers has both a Division of Waterway, Port, Coastal, and Ocean Engineering and a Coastal Engineering Research Council; several texts on coastal engineering and related specialized topics such as coastal hydraulics, dredging, port and harbor engineering, and beach processes have been published; and several universities have graduate courses and entire graduate programs in coastal engineering, usually through their Department of Civil Engineering.*

1.1. COASTAL ENGINEERING ACTIVITIES

Areas of concern to coastal engineers are demonstrated by the following list of typical coastal engineering activities.

- Development (through measurements and hindcasts) of nearshore wave, current, and water level design conditions.
- Design of a variety of stable and economic coastal structures including breakwaters, jetties, groins, seawalls, revetments, piers, towers, and pipelines.
- Control of beach erosion by the construction of coastal structures and/or by the artificial nourishment of beaches.
- Stabilization of tidal entrances by dredging, construction of structures, and mechanically bypassing sediment transported alongshore by waves and currents.
- Prediction of inlet and estuary currents and water levels and their effect on water quality, sediment movement, navigation, and so on.
- Effective disposal of cooling water and other liquid waste through marine pipeline outfalls in estuaries and the nearshore zone.
- Control and collection of spilled oil.
- Development of works to protect coastal areas from inundation by storm surge and tsunamis.
- Design of harbors and their appurtenances including quays, bulkheads, dolphins, wharves, and mooring systems.
- Functional and structural design of offshore islands.
- Dredging for adequate navigation and the subsequent effective disposal of dredged materials.

The solution of coastal engineering problems requires an understanding of basic ocean and coastal zone phenomena. Efforts to achieve this understanding may be classified in three categories.

*The report "University Curricula in the Marine Sciences and Related Fields" published by the Sea Grant Office of the National Ocean and Atmospheric Administration, provides an excellent summary of programs.

Theoretical and Mathematical Developments

Coastal engineers rely heavily on analytical theories for such things as the calculation of wave characteristics, the prediction of pressure distributions and total forces on structures owing to wave motion, and the prediction of waste diffusion in a turbulent flow field. During the past decade important advances have been made in the computer solution of the equations of continuity and motion written in finite difference form. These solutions help to investigate wave propagation problems such as tides in estuaries, refraction of waves into shallow water, and travel of tsunamis across the deep ocean and onto the continental shelves. These numerical models have also been used to study the rise and fall of water levels owing to storm surge and tide-generated flows through coastal inlets.

Laboratory Investigations

Experiments in wave and current flumes—investigating basic interactions of waves, wind, and currents with structures and sediment, as well as hydraulic model studies of specific problems at given locations—have been of immense value in advancing the art and science of coastal engineering. Examples include model studies of wind wave propagation into harbors; estuary models to investigate effects of dredging and other improvements on salt-water intrusion, current patterns, sediment deposition, and storm surge flooding; and coastal models with simulated movable bed material to evaluate sedimentary erosion and deposition patterns.

Field Studies

Problems that cannot be solved by analytical or laboratory means must be investigated through the collection of data in the field. The current understanding of phenomena such as the generation of waves by wind, response of beaches to wave attack, and the effects of coastal structures on beach processes is due largely to field measurements at many locations over long periods of time. There is a need for greater post-construction monitoring efforts at most types of coastal works.

1.2.   COASTAL ENGINEERING LITERATURE

This text does not serve as a coastal engineering design manual and it does not provide a survey of the applied aspects of coastal engineering. For coverage in these areas see Wiegel (1964), U.S. Army Corps of Engineers (1973), and Silvester (1974). However, references for each chapter were selected partially on the basis of availability and general subject coverage, with the expectation that the reader will make generous use of them. This use will expand the coverage of topics presented and will provide the reader with a greater familiarity with coastal engineering information sources. A selected list of the more common references in the coastal engineering literature follows.

Textbooks

Barber, N. F. (1969), *Water Waves*, Wykeham Publications, 142 p.

Brahtz, J. F. (1968), *Ocean Engineering*, John Wiley and Sons, 720 p.

Bretschneider, C. L., *Topics in Ocean Engineering*, Gulf Publishing Co., Vol. 1, 1969, Vol. 2, 1970.

Bruun, P. (1976), *Port Engineering*, Gulf Publishing Co., 586 p.

Cornick, H. F. (1962), *Dock and Harbor Engineering*, 4 Vols., Charles Griffen and Co., 1450 p.

Davies, J. L. (1973), *Geographical Variation in Coastal Development*, Hafner Publishing Co., 204 p.

Herbich, J. B. (1975), *Coastal and Deep Ocean Dredging*, Gulf Publishing Co., 622 p.

Houston, J. (1970), *Hydraulic Dredging*, Cornell Maritime Press, 318 p.

Ippen, A. T. (1966), *Estuary and Coastline Hydrodynamics*, McGraw-Hill, 744 p.

King, C. A. M. (1972), *Beaches and Coasts*, Edward Arnold, 570 p.

Kinsman, B. (1965), *Wind Waves*, Prentice-Hall, 676 p.

Komar, P. D. (1976), *Beach Processes and Sedimentation*, Prentice-Hall, 429 p.

LeMehaute, B. (1976), *An Introduction to Hydrodynamics and Water Waves*, Springer-Verlag, 315 p.

McCormick, M. E. (1973), *Ocean Engineering Wave Mechanics*, John Wiley and Sons, 179 p.

Minikin, R. R. (1963), *Winds, Waves and Maritime Structures*, Charles Griffen and Co., 294 p.

Muga, B. J. and J. F. Wilson (1970), *Dynamic Analysis of Ocean Structures*, Plenum Press, 377 p.

Muir Wood, A. M. (1968), *Coastal Hydraulics*, Gordon and Breach Science Publ., 187 p.

Myers, J. J., C. H. Holm and R. F. McAllister (1969), *Handbook of Ocean and Underwater Engineering*, McGraw-Hill, 1100 p.

Neumann, G. and W. J. Pierson (1966), *Principles of Physical Oceanography*, Prentice-Hall, 545 p. (one of several good physical oceanography texts available).

Quinn, A. D. (1972), *Design and Construction of Ports and Marine Structures*, McGraw-Hill, 661 p.

Silvester, R. (1974), *Coastal Engineering I, II*, Elsevier Scientific Publishing Co., 795 p.

U.S. Army Coastal Engineering Research Center (1973), *Shore Protection Manual*, Government Printing Office, 3 Vols.

Wiegel, R. L. (1964), *Oceanographical Engineering*, Prentice-Hall, 532 p.

Conference Proceedings

Conference on Coastal Engineering, American Society of Civil Engineers; First Conference, 1950, through Fifteenth Conference, 1976, plus occasional specialty conferences.

Offshore Technology Conference, Houston, Texas, annually since 1969.

World Dredging Conference, World Dredging Association, every 18 months since 1967.

Magazines and Journals

Bulletin, Permanent International Association of Navigation Congresses, Brussels, Belgium, published quarterly.

The Dock and Harbour Authority, Foxlow Publishing Co., London, published monthly.

Journal, Marine Technology Society, Washington, D.C., published ten times per year.

Ocean Engineering, Pergamon Press, Elmsford, N. Y.

Ocean Industry, Gulf Publishing Co., Houston, published monthly.

Shore and Beach, Journal of the American Shore and Beach Preservation Association, Miami, published semi-annually.

World Dredging and Marine Construction, Symcon Publishing Co., San Pedro, California, published monthly.

Proceedings, American Society of Civil Engineers, Waterway, Port, Coastal, and Ocean Engineering Division, published four times per year.

Government Publications

U.S. Army Coastal Engineering Research Center, Technical Memoranda, Special Reports (formerly Beach Erosion Board, BEB T.M. 1-135).

U.S. Army Waterways Experiment Station, Technical Reports, Research Reports, Miscellaneous Papers.

U.S. Army Corps of Engineers District Reports on particular projects and hurricane surveys.

U.S. Naval Civil Engineering Laboratory, Technical Reports, Technical Notes.

University Reports

A number of universities conduct research on coastal engineering and publish reports that have limited circulation.

1.3.   INTERNATIONAL SYSTEM OF MEASURING UNITS (SI)

Although the metric system of units has been extensively utilized in the fields of pure and applied science, the engineering profession has resisted its use until recently, preferring the English Gravitational System (pounds, slugs, feet, etc.). In time, a modernized version of the metric system, called the International System of Measuring Units or SI will be the standard system in the United States and will be the system used by the engineering profession. Leffel (1976) discusses use of the SI system by civil engineers as well as the basic components of the SI system and sources of information on this system. Table 1.1 gives conversion factors for English Gravitational to SI units for some of the basic units most commonly used by coastal engineers.

TABLE 1.1.   UNIT CONVERSION FACTORS

Multiply	By	To Obtain
Inch	2.540	centimeters
Feet	0.305	meters
Foot-pounds (energy)	1.356	joules
Foot-pounds/second (power)	1.356	watts
Gallons	3.785	liters
Horsepower	745.700	watts
Miles	1.609	kilometers
Miles/hour	0.868	knots
Pounds	4.448	newtons
Pounds/square foot (pressure, stress)	47.878	pascals
Sec^{-1}	1.000	hertz
Slugs	14.594	kilograms

1.4.   REFERENCES

Bruun, P. (1972), "The History and Philosophy of Coastal Protection" *Proceedings, Thirteenth Conference on Coastal Engineering*, American Society of Civil Engineers, pp. 33–74.

Inman, D. L. (1974), "Ancient and Modern Harbors: A Repeating Phylogeny" *Proceedings, Fourteenth Conference on Coastal Engineering*, American Society of Civil Engineers, pp. 2049–2067.

Leffel, R. E. (1976), "Civil Engineering Calculations Using SI Units" *Engineering Issues-Journal, Professional Activities*, American Society of Civil Engineers, January, pp. 99–116.

Silvester, R. (1974), *Coastal Engineering I, II*, Elsevier Scientific, N.Y., 795 p.

U.S. Army Coastal Engineering Research Center (1973), *Shore Protection Manual*, 3 vols., U.S. Government Printing Office, Washington, D.C.

Wiegel, R. L. (1964), *Oceanographical Engineering*, Prentice Hall, Englewood Cliffs, N. J., 532 p.

Two

TWO-DIMENSIONAL
WAVE EQUATIONS
AND WAVE CHARACTERISTICS

The most important environmental phenomenon with which a coastal engineer must contend is wave action in its various forms (e.g. wind waves, ship-generated waves, tsunamis, and the tide). Typically of less importance are the wind, storm surges, currents, and ice. These phenomena are, of course, interrelated in a number of ways. For example, wind-generated waves that approach the shore at an angle to the shoreline generate an alongshore current in the surf zone. The tide generates significant currents where its motion is restricted, such as at the entrance to bays and estuaries.

Waves at the air-sea interface transfer energy from the source that generates the waves to some structure or shoreline that dissipates and/or reflects the wave energy. Water particles effected by wave motion move in essentially closed orbits accompanied by only a small transport of mass in the direction of wave motion.

Figure 2.1 presents a wave classification system (wave energy spectrum) based on the estimated relative levels of wave energy for the range of surface wave periods found on the ocean (and on many other large water bodies). Also indicated are the primary wave generating forces for the various regions of this wave energy spectrum. Of greatest interest to coastal engineers are the wind-generated waves having periods typically between 2–20 sec. Wind-generated waves may be classified as either sea or swell. The former are waves

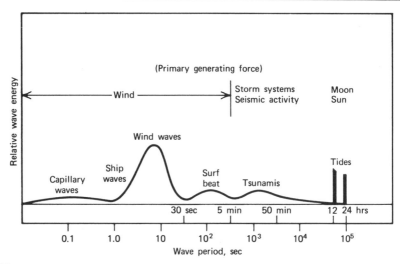

Figure 2.1 Estimated relative ocean wave energy and primary generating forces (adapted from Munk, 1950).

still under the influence of the generating wind. Swell are waves that propagate across the surface free of significant influence from the wind.

Also of engineering importance are tides, tsunamis (seismically generated sea waves), and waves generated by ships. Note that the spectrum is essentially continuous over the range of periods from capillary waves to the tides. Although all wave periods will not be present at a given place and time, usually many different periods will exist, even if only at low energy levels. For example, a detailed analysis of the time-history of the water surface elevation at a point in Lake Michigan might show wind waves with periods ranging from 2–6 sec, lake oscillations generated by a moving wind system and having fairly discrete periods ranging from a fraction of an hour to several hours, and a tide with component periods around 12 and 24 hr.

Waves at sea are very complex in that successive wave phases in a train of waves have different periods and amplitudes, with shorter and lower waves superimposed on longer and higher waves. However, as a starting point, and for certain research and design purposes, it is extremely valuable to have a simple theory describing a train of uniform period and amplitude (monochromatic) waves moving in water of constant depth. Several theories exist, varying both in degree of complexity and in accuracy for describing real waves. The simplest and generally most useful theory (considering the effort required in its use) is the small amplitude or linear wave theory first presented by Airy (1845). This theory provides equations for most of the kinematic and dynamic properties of surface gravity waves and predicts these properties within useful limits in most practical circumstances. The assumptions required, an outline of the derivation of the small-amplitude theory, the pertinent equations that result, and the important characteristics of waves described by these equations will be presented in this chapter. More detail on the Airy and other wave theories may be found in publications by Wiegel (1964), Kinsman (1965), Ippen (1966), and McCormick (1973).

2.1. SMALL AMPLITUDE WAVE THEORY

The small amplitude theory for two-dimensional, free, periodic gravity waves is developed by linearizing the equation that defines the free surface boundary condition. With this and the bottom boundary condition, a periodic velocity potential is sought that satisfies the requirements for irrotational flow. This velocity potential is then used to derive the equations for various wave characteristics (e.g. wave celerity, particle velocity and acceleration, pressure). Specifically, the assumptions required in this development are:

1. The water is homogeneous, incompressible, and surface tension forces are negligible. Thus there are no internal gravity or pressure waves affecting the flow, and the surface waves being considered are larger than that size where capillary effects are important (i.e. wave lengths greater than about 3 cm).

2. Flow is irrotational. Thus, there is no shear stress at the air-sea interface or at the bottom. Waves under the effects of wind (being generated or diminished) are not considered and the fluid slips freely at the bottom and other solid surfaces. Thus the velocity potential ϕ must exist and satisfy the Laplace equation

$$\frac{\partial^2 \phi}{\partial x^2} + \frac{\partial^2 \phi}{\partial y^2} = 0 \tag{2.1}$$

which is an expression of continuity for irrotational flow.

3. The bottom is not moving and is impermeable and horizontal. Therefore, the bottom is not adding or removing energy from the flow or reflecting wave energy. Waves shoaling over a sloping bottom can be defined, if the slope is small, by the theories developed later.

4. The pressure along the air-sea interface is constant. Thus no wind pressures exist and hydrostatic pressure differences due to surface elevation differences are negligible.

5. The wave amplitude is small compared to the wave length and water depth. Since particle velocities are related to wave amplitude, and the wave celerity (phase velocity) is related to the fluid depth and wave length, this implies that particle velocities are small compared to the wave celerity. This assumption, the most restrictive, allows one to linearize the higher order free surface boundary equation to obtain an easier solution. Higher order wave theories relax this assumption to a greater extent. This assumption means the linear wave theory is very limited in shallow water and near breaking, where waves peak and crest particle velocities approach the phase celerity.

Figure 2.2 shows a surface monochromatic wave moving at a celerity C in water of depth d in an x, y coordinate system. The wave height H (twice the amplitude) and length L are as shown. The wave period T is related to the wave length and celerity by $C = L/T$. The position of a particle at an instant during its orbital motion is given by horizontal and vertical coordinates ζ and ε respectively, referenced to the center of the orbit. The particle velocity components at any instant are u and v and the water surface elevation above the stillwater level (x-axis) at any point is η. Assume the water surface profile to be given as a function of position and time t by

$$\eta = \frac{H}{2} \cos 2\pi \left(\frac{x}{L} - \frac{t}{T} \right)$$

or

$$\eta = \frac{H}{2} \cos \left(kx - \sigma t \right) \tag{2.2}$$

where

$$k = \frac{2\pi}{L} \, (\text{wave number})$$

$$\sigma = \frac{2\pi}{T} \, (\text{wave angular frequency})$$

The arrows at the wave crest, trough, and stillwater lines indicate the directions of water particle motion at the surface. These motions cause a water particle to move in a clockwise orbit as the wave progresses from left

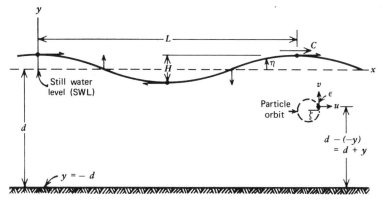

Figure 2.2 Definition sketch for progressive surface wave.

to right. The water particle velocities and orbit sizes decrease with increasing depth. Orbits are circular only under certain conditions as defined in section 2.3.

At the bottom there is no flow normal to the bed, which yields the boundary condition

$$v = \frac{\partial \phi}{\partial y} = 0 \bigg| y = -d \qquad (2.3)$$

The unsteady Bernoulli equation for irrotational flow may be written

$$\frac{1}{2}(u^2 + v^2) + gy + \frac{p}{\rho} + \frac{\partial \phi}{\partial t} = 0 \qquad (2.4)$$

where g is the acceleration of gravity, p is the pressure, and ρ is the fluid density. If one linearizes Eq. 2.4 by letting the velocity terms be zero and then applies the equation to the surface where the pressure is zero, one obtains

$$y = \eta = -\frac{1}{g}\frac{\partial \phi}{\partial t}\bigg| y = \eta \qquad (2.5)$$

Equation 2.5 defines the water surface boundary condition, which is approximately equal to the condition at the stillwater line if the wave amplitude is small, as required by the linearizing process. Thus the second boundary condition becomes

$$\eta = -\frac{1}{g}\frac{\partial \phi}{\partial t}\bigg| y = 0 \qquad (2.6)$$

Since the velocity potential should be cyclic with horizontal position and time and should vary with depth one can assume

$$\phi = Y\sin(kx - \sigma t) \qquad (2.7)$$

where $Y = f(y)$ only. Substitution of ϕ (Eq. 2.7) into the Laplace equation (Eq. 2.1) yields

$$\frac{\partial^2 Y}{\partial y^2} - k^2 y = 0$$

The general solution to this partial differential equation (see Kreyszig, 1962 for example) is

$$Y = Ae^{ky} + Be^{-ky}$$

where A and B are arbitrary constants. Thus from Eq. 2.7

$$\phi = (Ae^{ky} + Be^{-ky})\sin(kx - \sigma t) \qquad (2.8)$$

which satisfies the Laplace equation. Equation 2.8 will now be made to satisfy the boundary conditions (Eq. 2.3 and 2.6) in order to evaluate the constants A and B.

From Eq. 2.3

$$\frac{\partial \phi}{\partial y} = k(Ae^{-kd} - Be^{kd})\sin(kx - \sigma t) = 0$$

Since neither k nor $\sin(kx - \sigma t)$ are always zero

$$Ae^{-kd} - Be^{kd} = 0$$

or

$$A = \frac{Be^{kd}}{e^{-kd}}$$

and with some manipulation

$$\phi = Be^{kd}\left[e^{k(d+y)} + e^{-k(d+y)} \right]\sin(kx - \sigma t)$$

As the term in brackets equals $2\cosh k(d+y)$, the expression for the velocity potential can be written

$$\phi = 2Be^{kd}\cosh k(d+y)\sin(kx - \sigma t) \qquad (2.9)$$

From the second boundary condition (Eq. 2.6)

$$\eta = -\frac{1}{g}\frac{\partial\phi}{\partial t} = \frac{H}{2} \text{ at } t=0, x=0, y=0$$

Thus

$$\frac{gH}{2} = -\frac{\partial\phi}{\partial t} = 2\sigma B e^{kd}\cosh(kd)\cos(kx-\sigma t)$$

where $\cos(kx-\sigma t)=1$ at a wave crest. So

$$2Be^{kd} = \frac{Hg}{2\sigma\cosh kd}$$

and Eq. 2.9 becomes

$$\phi = \frac{H}{2}\frac{g\cosh k(d+y)}{\sigma\cosh kd}\sin(kx-\sigma t) \qquad (2.10)$$

which is the velocity potential we have been seeking.

The vertical component of velocity of a particle on the water surface v is given by $v=\partial\eta/\partial t$ where η is given by Eq. 2.6. Thus

$$v = -\frac{1}{g}\frac{\partial^2\phi}{\partial t^2}\bigg|_{y=0}$$

Also, since

$$v = \frac{\partial\phi}{\partial y}$$

the previous expression can be written as

$$\frac{\partial^2\phi}{\partial t^2} + g\frac{\partial\phi}{\partial y} = 0$$

Inserting ϕ from Eq. 2.10 and solving yields

$$\sigma^2 = gk\tanh(kd) \qquad (2.11)$$

Since

$$\frac{\sigma}{k} = \frac{L}{T} = C$$

then

$$C = \sqrt{\frac{gL}{2\pi} \tanh \frac{2\pi d}{L}} \tag{2.12}$$

Equation 2.12 is the basic equation that relates the wave celerity, the wave length, and the water depth. Note that celerity is independent of wave height according to the small amplitude wave theory. Equation 2.12 can also be written

$$C = \frac{gT}{2\pi} \tanh\left(\frac{2\pi d}{L}\right) \tag{2.13}$$

and

$$L = \frac{gT^2}{2\pi} \tanh\left(\frac{2\pi d}{L}\right) \tag{2.14}$$

As a wave travels from deep water toward the shore its length, celerity, height, surface profile, and internal pressure and velocity fields vary. The wave period, however, remains constant.

2.2. WAVE CLASSIFICATION BY RELATIVE DEPTH

An important wave classification system based upon the relative water depth (d/L) is demonstrated by the relationships shown in Fig. 2.3*. When the relative depth is greater than approximately 0.5, $\tanh(2\pi d/L) \cong 1.0$ and Eqs. 2.11–2.14 reduce to

$$C_0 = \sqrt{\frac{gL_0}{2\pi}} \tag{2.15}$$

$$C_0 = \frac{gT}{2\pi} \tag{2.16}$$

and

$$L_0 = \frac{gT^2}{2\pi} \tag{2.17}$$

---

*Extensive tables of these and other functions can be found in Wiegel (1964) and U.S. Army Coastal Engineering Research Center (1973).

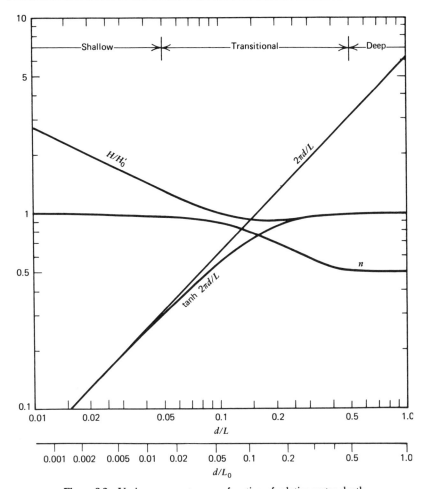

Figure 2.3 Various parameters as a function of relative water depth.

This condition is called deep water and is typically denoted by the subscript zero. In deep water, partical orbital motion decays with depth so it is near zero at $-y/L > 0.5$. Wave characteristics are independent of the water depth as indicated by Eqs. 2.15–2.17. As Fig. 2.3 demonstrates, there is only a few percent error in assuming deep water conditions apply for depths greater than a third of the wave length.

In order to demonstrate orders of magnitude, consider a typical deep water wave having a 10-sec period and a height of 2 m. From Eqs. 2.16 and 2.17 the celerity would be 15.6 m/s (34.9 mph.) and the length would be

156 m. The wave steepness (H_0/L_0) would be 0.013 and the speed of a water particle at the surface (orbit circumference/wave period $= \pi H / T$) would be 0.63 m/s. The low wave steepness and particle velocity (in comparison to wave celerity) should be noted.

When the relative depth is less than approximately 0.05, $\tanh(2\pi d/L) \cong 2\pi d/L$ and Eq. 2.12 becomes*

$$c = \sqrt{gd} \qquad (2.18)$$

Wave celerity is now dependent only on the water depth, and the wave length and period are related by

$$L = \sqrt{gd}\ T$$

This condition is called shallow water. Remember, it is the relative depth that is important. For example, the tide is a very long wave that behaves as a shallow water wave even in the deepest regions of the ocean. Continuing our example, a 10-sec wave in 2 m of water would have a length of 44.3 m (thus $d/L < 0.05$) and a celerity of 4.4 m/s.

For $0.5 > d/L > 0.05$ the transitional condition exists and Eqs. 2.12–2.14 must be used. Dividing Eq. 2.13 by Eq. 2.16, or Eq. 2.14 by Eq. 2.17 we have

$$\frac{C}{C_0} = \frac{L}{L_0} = \tanh\left(\frac{2\pi d}{L}\right) \qquad (2.19)$$

which, when compared with Fig. 2.3, indicates the decrease in wave celerity and length that occurs as a wave travels toward shore. Further manipulation of Eq. 2.19 yields

$$\frac{d}{L}\tanh\left(\frac{2\pi d}{L}\right) = \frac{d}{L_0} \qquad (2.20)$$

which allows us to calculate the wave length for any water depth, given the deep water wave length. Values of d/L_0 as a function of d/L are shown on Fig. 2.3.

---

*Derivation of the small amplitude wave theory assumes the amplitude is small compared to the length, which is not always valid in shallow water. Finite amplitude wave theories indicate that d in Eq. 2.18 is best replaced by the depth plus wave crest amplitude when the amplitude is significant compared to the depth.

2.3. WAVE KINEMATICS AND PRESSURE

The horizontal and vertical components of water particle velocity (u and v) may be determined from the velocity potential where $u = \partial\phi/\partial x$ and $v = \partial\phi/\partial y$. After some algebraic manipulation, we have

$$u = \left(\frac{\pi H}{T}\right)\left(\frac{\cosh k(d+y)}{\sinh kd}\right)\cos(kx - \sigma t) \qquad (2.21)$$

and

$$v = \left(\frac{\pi H}{T}\right)\left(\frac{\sinh k(d+y)}{\sinh kd}\right)\sin(kx - \sigma t) \qquad (2.22)$$

Note that each velocity component is defined by a term consisting of three parts: (1) the surface deep water particle speed, $\pi H/T$, (2) a hyperbolic term defining the modification of particle speed with increasing relative water depth and particle position along a vertical axis, and (3) a phasing term dependent on position and time. The distance $d + y$ is measured from the bottom to the particle location as demonstrated in Fig. 2.2.

The horizontal component of acceleration of a particle is given by

$$a_x = \underbrace{u\frac{\partial u}{\partial x} + v\frac{\partial u}{\partial y}}_{\text{convective}} + \underbrace{\frac{\partial u}{\partial t}}_{\text{local}}$$

However, a consequence of the small amplitude assumptions is that the convective accelerations are usually negligible. Thus

$$a_x \cong \frac{\partial u}{\partial t} = \frac{2\pi^2 H}{T^2}\frac{\cosh k(d+y)}{\sinh kd}\sin(kx - \sigma t) \qquad (2.23)$$

In a similar way

$$a_y \cong \frac{\partial v}{\partial t} = -\frac{2\pi^2 H}{T^2}\frac{\sinh k(d+y)}{\sinh kd}\cos(kx - \sigma t) \qquad (2.24)$$

The equations for particle acceleration have the same depth decay term as the velocity equations and show the particle acceleration to be 90° out of phase with the particle velocity.

As water particles orbit around a mean position (see Fig. 2.2) the horizontal and vertical ordinates of particle displacement (ζ and ε) are related to the components of particle velocity by $u = \partial\zeta/\partial t$ and $v = \partial\varepsilon/\partial t$. Thus

$$\zeta = \int u\,dt = \frac{H}{2}\frac{\cosh k(d+y)}{\sinh kd}\sin(kx - \sigma t)$$

and

$$\varepsilon = \int v\,dt = \frac{H}{2}\frac{\sinh k(d+y)}{\sinh kd}\cos(kx - \sigma t)$$

where $H/2$ is the orbit radius for a surface particle with deep water wave motion. As a wave travels from deep to shallow water, the particle orbits undergo the transformation demonstrated in Fig. 2.4.

Orbits change from being circular throughout the water column in deep water to being elliptical, with the ellipses becoming flatter with depth, in transitional and shallow water. Also, in transitional and shallow water the particle motion at the bottom is strictly horizontal. The bottom velocity given by Eq. 2.21 applies near, but not at, the bottom as bed shear stresses develop a boundary layer in which the horizontal velocity component reduces to zero at the bottom.

The water surface profile becomes trochoidal with long flat troughs and shorter peaked crests. Also, the amplitude of the wave crest increases and the amplitude of the trough decreases with the total wave height being greater than the deep water height. In transitional and shallow water, particles still move in essentially closed orbits. Since they must travel the same distance in less time under the crest owing to the trochoidal surface profile, crest velocities will be greater than trough velocities.

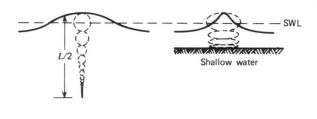

Figure 2.4 Wave transformation, deep to shallow water (vertical scale exaggerated).

For deep and shallow water the following relationships are useful in the calculation of particle velocities, accelerations and orbit displacements.

$$\text{Deep water:} \qquad \frac{\cosh k(d+y)}{\sinh kd} \approx \frac{\sinh k(d+y)}{\sinh kd} \approx e^{ky} \qquad (2.27)$$

$$\text{Shallow water:} \quad \frac{\cosh k(d+y)}{\sinh kd} \approx \frac{1}{kd} \qquad (2.28)$$

$$\frac{\sinh k(d+y)}{\sinh kd} \approx 1 + \frac{y}{d} \qquad (2.29)$$

Equation 2.27 indicates that deep water particle velocities, accelerations, and orbit diameters decay exponentially with depth. Substitution of Eqs. 2.28 and 2.29 into Eqs. 2.21 and 2.22 and minor algebraic manipulation leads to the following equations for shallow water waves.

$$u = \frac{H}{2}\sqrt{\frac{g}{d}} \; \cos(kx - \sigma t) \qquad (2.30)$$

$$v = \frac{\pi H}{T}\left(1 + \frac{y}{d}\right)\sin(kx - \sigma t) \qquad (2.31)$$

As Eqs. 2.30 and 2.31 indicate, in shallow water the horizontal component of particle velocity is independent of distance below the stillwater level and the vertical component diminishes from a maximum at the surface to zero at the bottom. Similar equations can be written for particle displacement.

Substitution of the velocity potential (Eq. 2.10) into the linearized form of the equation of motion (Eq. 2.4) yields the following equation for the pressure

$$p = -\rho g y + \frac{\rho g H}{2}\frac{\cosh k(d+y)}{\cosh kd} \cos(kx - \sigma t) \qquad (2.32)$$

The first term on the right is the normal hydrostatic pressure and the second term is the dynamic pressure due to particle acceleration. These components are plotted in Fig. 2.5 for vertical sections through the wave crest and trough. Since water particles under the wave crest are accelerating downward, a downward dynamic pressure gradient is required. The reverse occurs under a wave trough. Halfway between the crest and trough the acceleration is horizontal so the vertical pressure distribution is hydrostatic. Eq. 2.32 is

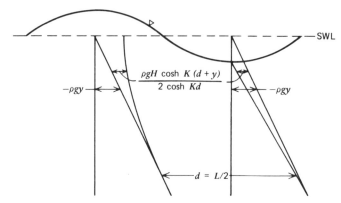

Figure 2.5 Vertical pressure distribution, deep water wave (vertical scale exaggerated).

not valid above the stillwater line as a consequence of the surface boundary condition being applied at $y = 0$ (Eq. 2.6).

An underwater pressure sensing device located at a depth of less than a half wave length would detect the dynamic pressure fluctuation and allow computation of the wave height and period (prob. 3). This is one of the more common techniques used to measure waves in coastal regions.

2.4. WAVE ENERGY AND POWER

The total energy in a wave is the sum of the kinetic and potential energies. Equations for each may be derived individually by consulting Fig. 2.6. The kinetic energy per unit width of wave crest and for one wave length E_k is equal to the integral over one wave length and the water depth of one-half times the mass of a differential element times the velocity of that element squared. Thus

$$E_k = \int_0^L \int_{-d}^0 \frac{1}{2} \rho \, dx dy (u^2 + v^2)$$

Inserting Eqs. 2.21 and 2.22 for the velocity components, integrating, and performing the required algebraic manipulation yields

$$E_k = \frac{\rho g H^2 L}{16}$$

If we subtract the potential energy of a mass of stillwater from the potential energy of the wave form shown in Fig. 2.6, we will have the

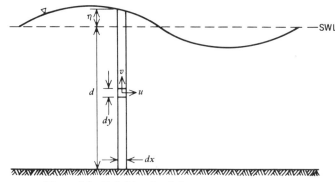

Figure 2.6 Definition sketch for wave energy derivation.

potential energy due solely to wave motion. Using the bottom as a datum plane, the potential energy per unit crest width in one wave length E_p is

$$E_p = \int_0^L \rho g (d + \eta) \left(\frac{d + \eta}{2} \right) dx - \rho g L d \left(\frac{d}{2} \right)$$

The surface elevation as a function of x is given by Eq. 2.2 (Let $t = 0$). Performing the calculus, we have

$$E_p = \frac{\rho g H^2 L}{16}$$

Thus the kinetic and potential energies are equal and the total energy in a wave per unit crest width E is

$$E = E_k + E_p = \frac{\rho g H^2 L}{8} \tag{2.33}$$

A wave propagating through a porous structure, for example, where the water depth is the same on both sides, will have the same wave length on both sides (see Eq. 2.14). Thus a reduction in wave energy because of reflection and viscous dissipation will result in a decrease in wave height. A 50 percent reduction in wave energy would only result in a 29 percent decrease in wave height since the wave energy is proportional to the wave height squared.

The energy is variable from point to point along a wave length but the average energy per unit surface area is

$$\bar{E} = \frac{E}{L} = \frac{\rho g H^2}{8} \tag{2.34}$$

This is occasionally referred to as the energy density or specific energy.

Wave power is the wave energy per unit time propagated in the direction of wave travel. The power can be written as the product of the force acting on a vertical plane normal to wave propagation times the particle flow velocity across this plane. For a unit crest width, the average wave power is

$$P = \frac{1}{T} \int_0^T \int_{-d}^0 (p + \rho g y) u \, dt \, dy$$

Here, the force is the dynamic pressure $p + \rho g y$ times a unit surface area $1(dy)$. Inserting the dynamic pressure from Eq. 2.32 and the horizontal component of velocity from Eq. 2.21 and integrating yields

$$P = \frac{\rho g H^2 L}{16 T} \left(1 + \frac{2kd}{\sinh 2kd} \right) \tag{2.35}$$

or

$$P = \frac{E}{T} \left[\frac{1}{2} \left(1 + \frac{2kd}{\sinh 2kd} \right) \right] \tag{2.36}$$

Letting

$$n = \frac{1}{2} \left(1 + \frac{2kd}{\sinh 2kd} \right) \tag{2.37}$$

Eq. 2.36 becomes

$$P = \frac{nE}{T} \tag{2.38}$$

The functional relationship between the term n and the relative depth is indicated by Fig. 2.3, where n increases from 0.5 in deep water to 1.0 in shallow water. Equation 2.38 shows n can be thought of as the fraction of the wave energy that is transmitted forward in a wave.

As a wave train travels, the energy per unit time passing one point on its path must equal the energy per unit time passing a subsequent point plus the energy reflected or dissipated per unit time between the two points. If the rate of energy reflection and dissipation are negligible, a valid assumption in many cases,

$$P = \left(\frac{nE}{T} \right)_1 = \left(\frac{nE}{T} \right)_2 = \text{constant} \tag{2.39}$$

Equation 2.39 indicates, for example, that as a wave train travels from deep to shallow water the energy in the train must decrease at a rate inversely proportional to the increase in n since the wave period is constant.

If we construct lines normal or orthogonal to the wave crests as a wave train advances, and assume that no energy propagates along wave crests (i.e. across orthogonal lines), the power contained between orthogonal lines can be considered constant. If the orthogonal spacing is denoted by B,

$$\left(\frac{BnE}{T}\right)_1 = \left(\frac{BnE}{T}\right)_2 = \text{constant}$$

Inserting the wave energy from Eq. 2.33 and reorganizing

$$\frac{H_1}{H_2} = \sqrt{\frac{n_2 L_2}{n_1 L_1}} \sqrt{\frac{B_2}{B_1}} \qquad (2.40)$$

The first term on the right represents the effects of shoaling and the second term represents the effects of orthogonal line convergence or divergence due to wave refraction (Chapter 3). These are often called the coefficients of shoaling K_S and refraction K_R respectively. The variation of wave height from deep water to some transitional or shallow water depth without the effects of refraction (i.e. $K_R = 1$) can be calculated from Eq. 2.40 and is shown in Fig. 2.3 by the curve labeled H/H_0'. Initially, as a wave shoals there is a slight decrease in wave height because n increases at a faster rate than L decreases (see Eq. 2.35). At relative depths less than approximately 0.1 the wave height is greater than H_0 and it continues to increase with shoaling until the wave becomes unstable and breaks.

The assumption of zero energy dissipation as a wave shoals can lead to significant error in predicting wave heights in regions where waves travel in transitional or shallow water depths for long distances. The unsteady boundary layer at the bottom develops viscous stresses that dissipate wave energy. Bottom irregularities, such as large scale ripples, generate vortices that also dissipate energy. The fluctuating bottom pressures cause a cyclic flow into and out of the bed (if permeable), resulting in a further dissipation of wave energy. Further discussion of these energy dissipation mechanisms can be found in Ippen (1966).

Experiments in wave tanks (Wiegel, 1950; Eagleson, 1956; and leMéhauté et al., 1968) have given some indication of the accuracy of the small-amplitude theory in predicting wave characteristics, particularly in transitional and shallow water depths where the assumptions of relatively low wave height and water particle velocity are less valid. In summary:

1. For most typical beach slopes the equations for wave celerity and length (Eqs. 2.13 and 2.14) are satisfactory up to the wave breaker zone.

2. For increasing beach slopes and steeper waves, the wave height predicted by Eq. 2.40 will be lower than the real wave height. The discrepancy increases as the relative depth decreases. As an example, on a 1:10 slope, for a relative depth of 0.1 and a deep water wave steepness of 0.02, the experimental wave height exceeded the calculated wave height by 15 percent.

3. The small-amplitude theory assumes a sinusoidal water surface profile quite different from the trochoidal profile a wave has in shallow water.

4. For waves on a flat slope and having a relative depth greater than about 0.1, the small-amplitude theory is satisfactory for predicting horizontal water particle velocities. At lesser relative depths the small-amplitude theory still predicts reasonably good values for horizontal velocity near the bottom, but results are poorer (up to 50 percent errors on the low side) near the surface. However, some of the more sophisticated wave theories were often less accurate than the small-amplitude theory at predicting water particle velocities in transitional and shallow water depths.

2.5. WAVE GROUP CELERITY

Consider a long wave tank in which a small group of waves is generated. As the waves travel along the tank, waves in the front of the group will gradually decrease in height and, if the tank is long enough, disappear in sequence starting with the first wave. As the waves in the front decrease in height, new waves will appear at the back and commence to grow. One new wave will appear each wave period so the total number of waves in the group will continually increase. This phenomenon means the wave group travels slower than individual waves in the group.

An explanation for this phenomenon can be found in the fact that as a wave travels forward only a fraction (n) of its energy propagates forward. Thus the waves at the front of a group diminish in amplitude. The last wave leaves energy behind so, relative to the group, a new wave appears each T seconds and gains additional energy as time passes. As the total energy in the group must remain constant (neglecting dissipation and reflection), the average amplitude of the group must continue to decrease as the number of waves increases with time.

A practical consequence of the group celerity being less than the celerity of individual waves is that when waves are generated by a storm, prediction of their arrival time at a point of interest must be based on the group celerity.

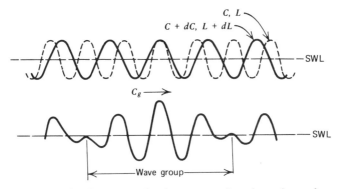

Figure 2.7 Two wave trains shown separately and superimposed.

In order to develop an equation for the group celerity let us consider two trains of monochromatic waves having slightly different periods. Figure 2.7 shows the wave trains both separately and superimposed, as if traveling in the same area. The superimposition of the two wave trains results in a beating effect in which the waves are alternately in and out of phase. This produces the highest wave where they are in phase, with heights diminishing in the forward and backward directions to zero height where the waves are exactly out of phase. The result is a wave group advancing at a celerity, C_g.

Referring to Fig. 2.7, the time required for the lag dL to be made up is dt, where dt is equal to the difference in wave lengths divided by the difference in wave celerity for the two wave trains (i.e. the distance to be caught up divided by the celerity at which the faster wave catches up). Thus

$$dt = \frac{dL}{dC}$$

The group advances a distance dx in the time dt, where dx is the distance traveled by the group in a time interval dt minus the one wave length that the peak wave dropped back (as the in-phase waves drop back one wave length each period). We can then write

$$dx = \left[\frac{(C+dC)+C}{2} \right] dt - \frac{(L+dL)+L}{2} \approx Cdt - L$$

and

$$C_g = \frac{dx}{dt} = \frac{Cdt - L}{dt} = C - \frac{L}{dt}$$

Since

$$dt = \frac{dL}{dC}$$

$$C_g = C - L\left(\frac{dC}{dL}\right) \tag{2.41}$$

Inserting Eq. 2.12 into Eq. 2.41 leads to

$$C_g = \frac{C}{2}\left(1 + \frac{2kd}{\sinh 2kd}\right) \tag{2.42}$$

Thus

$$C_g = nC \tag{2.43}$$

where the term n now also defines the ratio of the wave group to phase celerity. Remember, n varies from 0.5 in deep water to 1.0 in shallow water. Another way to look at this is that the wave energy is propagated at the group celerity.

2.6.   MASS TRANSPORT, WAVE SETUP

The small amplitude wave theory predicts that water particles move in closed orbits and, therefore, that there is no transport of mass as waves advance. However, in real waves there is a small forward mass transport as water particles advance slightly during each orbit. This mass transport velocity increases as the wave steepness and celerity increase and as the relative water depth decreases. Mass transport velocities are an order of magnitude lower than particle velocities but they are significant in causing a setup of water nearshore, with the subsequent generation of nearshore currents. In addition they will contribute to the transport of sediment placed into suspension by other mechanisms.

Based upon a laboratory study by Saville (1961) of waves shoaling on a beach, and upon theoretical work by Longuet-Higgins and Stewart (1963), an equation for the wave setup at the shore S_{ww} has been proposed by the U.S. Army Coastal Engineering Research Center (1973). If H_b is the breaker height in the surf zone,

$$S_{ww} = 0.19\left[1 - 2.82\sqrt{\frac{H_b}{gT^2}}\,\right]H_b \tag{2.44}$$

Typically, wave setup will amount to about 15 percent of the breaker height.

2.7. WAVE REFLECTION

If a wave train encounters a change in boundary conditions (e.g. change in bottom depth, convergence or expansion of wave tank walls, a submerged obstacle), a portion of the energy in the wave train will be reflected. If the obstacle is a vertical, frictionless, inelastic wall, the wave will completely reflect, resulting in a standing wave with an antinode height twice the incident wave height and with water particle motions as depicted in Fig. 2.8a. Particle orbits are flattened to produce trajectories with forward and return motion along the same curved lines. Particle motions vary from being vertical under the antinodal point to horizontal under the nodal point. The water surface is defined by the sum of an incident and a reflected wave or

$$\eta = \frac{H}{2}\cos(kx - \sigma t) + \frac{H}{2}\cos(kx + \sigma t) = H\cos(kx)\cos(\sigma t) \qquad (2.45)$$

Thus the standing wave height is $2H$.

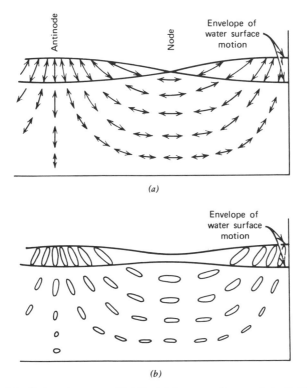

(a)

(b)

Figure 2.8   Reflected waves. a. $Hr/Hi = 1.0$ (pure standing wave); b. $Hr/Hi = 0.5$.

In a similar fashion, the velocity potential is the sum of the potentials for the incident and reflected waves or

$$\phi = -\frac{Hg}{\sigma}\frac{\cosh k(d+y)}{\cosh kd}\cos(kx)\sin(\sigma t) \qquad (2.46)$$

From the linearized form of Eq. 2.4 and Eq. 2.46

$$p = -\rho g y + \rho g H \frac{\cosh k(d+y)}{\cosh kd}\cos(kx)\cos(\sigma t) \qquad (2.47)$$

which gives the pressure distribution throughout the wave and along the wall (antinode, $x=0$).

As the wall slope decreases or the wall roughness or permeability increase, the height of the reflected wave decreases. Also, for a given obstacle, wave reflection decreases with an increase in incident wave steepness, particularly when wave breaking occurs. Figure 2.8b shows the water particle motion and water surface envelope when the reflected wave height H_r is half of the incident wave height H_i. Particles move in flat curved elliptical paths oriented essentially as particle trajectories in a pure standing wave. Compare this particle motion with that in Fig. 2.8a and with that for a progressive oscillatory wave (Fig. 2.3) to see the progression in particle motions that occurs as wave reflection varies.

2.8. WAVE BREAKING

It was demonstrated in Section 2.2 that the crest particle velocity is typically much lower than the wave celerity. In deep water the crest particle velocity is proportional to the wave height for a given wave period (see Eq. 2.21), so with increasing wave heights this velocity will eventually equal the wave celerity. At this point the wave will become unstable and break. Also, as a wave shoals, the increasing crest particle velocity becomes equal to the decreasing phase velocity causing the wave to break.

Miche (1944) determined the limiting condition for wave breaking in any water depth to be given by

$$\left(\frac{H}{L}\right)_{\max} = \frac{1}{7}\tanh\left(\frac{2\pi d}{L}\right) \qquad (2.48)$$

Danel (1952) performed laboratory experiments in a wave tank with a horizontal bottom and found this equation to be sufficiently accurate for engineering purposes.

In deep water, Eq. 2.48 reduces to

$$\left(\frac{H_0}{L_0}\right)_{max} = \frac{1}{7} \tag{2.49}$$

That is, when the deep water wave height becomes one-seventh of the wave length, the wave will break. This has important consequences, for example, in the nature of the wave spectrum generated during a storm. Growth of the short period waves in the spectrum is often limited by this criterion. In shallow water, we have

$$\left(\frac{H}{L}\right)_{max} = \frac{1}{7}\frac{2\pi d}{L}$$

or

$$\left(\frac{H}{d}\right)_{max} = 0.9 \tag{2.50}$$

Waves breaking on a sloping beach are generally classified into three categories (see Fig. 2.9):

Spilling

Plunging

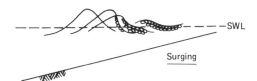

Surging

Figure 2.9   Wave breaking classification.

1. *Spilling.* Waves of relatively low steepness that shoal on flat slopes will break by a continuous spilling of foam down the front face of the wave until the wave is dissipated.

2. *Plunging.* With a relative increase in wave steepness and beach slope, a wave will peak and the crest will curl forward and plunge as the front face of the wave becomes concave.

3. *Surging.* At the steepest wave and slope conditions, a wave will peak like a plunging wave but before the crest plunges, the base of the wave will surge up the beach face.

Plate 2.1 Plunging breaker in large wave tank. (Courtesy of U.S. Army Coastal Engineering Research Center)

The type of breaker that occurs is often important to the stability of a structure subjected to breaking waves. It will also affect the amount of energy reflected from a slope and the elevation of wave runup on a slope.

Equation 2.50 gives the conditions for breaking of an ideal wave on a zero slope. Actual breaking conditions for a shoaling wave also depend on the beach slope and wave steepness (as well as the offshore bar shape and location, wind conditions, beach face permeability, and local currents). For shoaling waves it is best to predict breaking conditions using Figs. 2.10 and

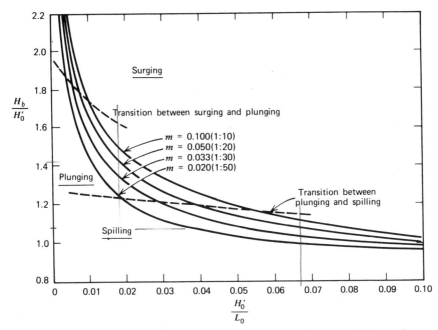

Figure 2.10 Breaker height and class as a function of slope, deep water height, and wave steepness (U.S. Army Coastal Engineering Research Center, 1973).

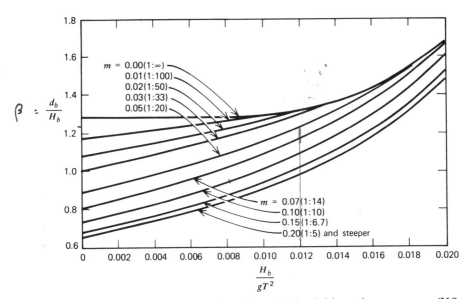

Figure 2.11 Breaker depth as a function of slope, breaker height, and wave steepness (U.S. Army Coastal Engineering Research Center, 1973).

2.11, which are based on experimental data and were presented by Goda (1970). Given the deep water unrefracted wave height, the wave period, and the beach slope, the breaker height can be determined from Fig. 2.10 and then the depth at breaking can be determined from Fig. 2.11. If a wave refracts upon shoaling, the hypothetical unrefracted wave height H_0' can be evaluated from

$$H_0' = K_R H_0 \qquad (2.51)$$

From laboratory observations, Wiegel and Beebe (1956) found that, at breaking, the ratio of the wave crest elevation above the stillwater level to the wave height was consistently about 0.78.

2.9. WAVE RUNUP

The elevation to which the crest of a coastal structure such as a seawall, stone revetment, and so on, is constructed depends, to a large extent, on the runup elevation and the allowable overtopping of the design wave. An understanding of wave runup is also important in the investigation of beach processes.

A large amount of laboratory wave runup data from project model studies and basic studies on plane and compound slopes is available. Saville (1957) has synthesized and presented much of this data in the form of Fig. 2.12, which gives the runup R (the vertical height above swl) as a function of the wave period, the unrefracted deep water height, and the cotangent of the structure slope. For refracted waves use the hypothetical unrefracted deep water height H_0' from Eq. 2.51. The curves are for smooth impermeable slopes with a water depth at the toe between one and three times H_0'. Curves for other relative depths are given in U.S. Army Coastal Engineering Research Center (1973). Figure 2.12 indicates that, for a given structure slope, steeper waves (higher H_0'/T^2) have a lower relative runup (R/H_0'). Also, for most beach and revetment slopes (slopes flatter than 1 on 2), the runup of a given wave increases as the slope becomes steeper.

Table 2.1 is a tabulation presented by Battjes (1970) of the effects of slope surface condition on wave runup based on a large number of laboratory studies in Europe and the United States. The factor r gives the ratio of runup on the given surface to that on a smooth impermeable surface and may be multiplied by the runup determined from Fig. 2.12 to predict the wave runup.

Saville (1957) suggested a procedure for applying Fig. 2.12 to composite slopes (see Fig. 2.13). A single hypothetical slope is constructed from the

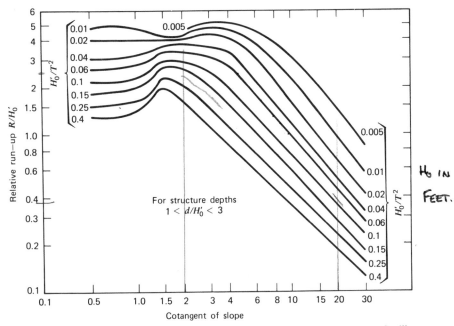

Figure 2.12 Wave runup as a function of wave characteristics and structure slope (Saville, 1957).

TABLE 2.1. (FROM BATTJES, 1970)

Slope Facing	r
Concrete slabs	0.9
Placed basalt blocks	0.85–0.9
Grass	0.85–0.9
One layer of riprap on an impermeable base	0.8
Placed stones	0.75–0.8
Round stones	0.6–0.65
Dumped stones	0.5–0.6
Two or more layers of riprap	0.5
Tetrapods, etc.	0.5

wave breaking point to the estimated point of wave runup on the composite slope. The runup on this hypothetical slope is then determined by trial and error from Fig. 2.12 and compared with the estimated runup on the composite slope to see if it agrees. If not, the procedure is continued until agreement is achieved. Saville reports good results from this approach when compared to laboratory data for composite slopes, provided any horizontal

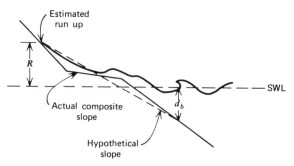

Figure 2.13   Wave runup on a composite slope (Saville, 1957).

or near horizontal berms on the slope near the still water level are not wider than about one-fourth of the incident wave length. For longer berms, the predicted runup will be low because water sets up on the berm and reformed waves or surges travel across the setup water and then run up the next slope.

Figure 2.12 is based on studies using wave trains of constant period and amplitude. However, the runup of a single wave in an irregular wave train can vary significantly from the value predicted by Fig. 2.12. For example, since the runup of one wave is affected by the downrush of the previous wave, a high and long wave preceeded by a relatively lower and shorter wave will run up much higher than if preceded by another high and long wave. Small waves following larger waves tend to be obliterated by the downrush so in a train of irregular waves there will typically be fewer runup crests than incident waves.

Also of importance in coastal engineering design is the volume of water from wave runup that will overtop a given structure crest elevation. The limited data available on this subject is presented by the U.S. Army Coastal Engineering Research Center (1973).

2.10. SUMMARY

This chapter has presented the small amplitude or Airy theory for two-dimensional waves. A familiarity with this theory and the two-dimensional characteristics of waves is indispensable to an understanding of much that follows. Chapter 3 deals with important three-dimensional aspects of waves including refraction, diffraction, and reflection. Chapter 5 covers several aspects of wind-generated waves, particularly their generation, measurement, analysis, and prediction. All the remaining chapters except Chapter 8 require some understanding of the mechanics of water waves.

2.11. REFERENCES

Airy, G. B. (1845), "On Tides and Waves," *Encyclopedia Metropolitana*, London, pp. 241–396.

Battjes, J. A. (1970), Discussion of "The Runup of Waves on Sloping Faces—A Review of the Present State of Knowledge," by N. B. Webber and G. N. Bullock, *Proceedings, Conference on Dynamics Waves in Civil Engineering*, Wiley, New York, pp. 293–314.

Danel, P. (1952), "On the Limiting Clapotis," *Gravity Waves*, National Bureau of Standards Circular 521, pp. 35–38.

Eagleson, P. S. (1956), "Properties of Shoaling Waves by Theory and Experiment," *Transactions, American Geophysical Union*, Vol. 37, pp. 565–572.

Goda, Y. (1970), "A Synthesis of Breaker Indices," *Transactions, Japanese Society of Civil Engineers*, Vol. 2, Part 2.

Ippen, A. T. (1966), *Estuary and Coastline Hydrodynamics*, McGraw-Hill, New York, 744 p.

Kinsman, B. (1965), *Wind Waves*, Prentice-Hall, Englewood Cliffs, N.J., 676 p.

Kreyszig, E. (1962), *Advanced Engineering Mathematics*, Wiley, New York, 856 p.

LeMéhauté, B., Divoky, D. and A. Lin (1968), "Shallow Water Waves: A Comparison of Theories and Experiments," *Proceedings, Eleventh Conference on Coastal Engineering*, American Society of Civil Engineers, pp. 86–107.

Longuet-Higgins, M. S. and R. W. Stewart (1963), "A Note on Wave Setup," *Journal of Marine Research*, Vol. 21, pp. 4–10.

McCormick, M. E. (1973), *Ocean Engineering Wave Mechanics*, Wiley, New York, 179 p.

Miche, R. (1944), "Movements ondulatoires des mers en profondeur constante ou décroissante," *Annales des Points et Chaussés*, pp. 25–78, 131–164, 270–292, 369–406.

Saville, T. (1957), "Wave Run-Up on Composite Slopes," *Proceedings, Sixth Conference on Coastal Engineering*, Council on Wave Research, University of California, Berkeley, pp. 691–699.

Saville, T. (1961), "Experimental Determination of Wave Setup," *Proceedings, Second Technical Conference on Hurricanes*, National Hurricane Research Project, Report No. 50, pp. 242–252.

U.S. Army Coastal Engineering Research Center (1973), *Shore Protection Manual*, 3 Vols., U.S. Government Printing Office, Washington, D.C.

Wiegel, R. L. (1950), "Experimental Study of Surface Waves in Shoaling Water," *Transactions, American Geophysical Union*, Vol. 31, pp. 377–385.

Wiegel, R. L. and K. E. Beebe (1956), "The Design Wave in Shallow Water," *Journal, Waterways and Harbors Division*, American Society of Civil Engineers, Vol. 82, Paper 910.

Wiegel, R. L. (1964), *Oceanographical Engineering*, Prentice-Hall, Englewood Cliffs, N.J., 532 p.

2.12. PROBLEMS

1. A wave tank at the U.S. Army Coastal Engineering Research Center is 193 m long, 4.57 m wide and 6.1 m deep. The tank is filled to a depth of 5 m with fresh water and a 1-m high, 4-sec period wave is generated.
   a. Calculate the wave celerity, length, group celerity, energy, and power.

b. Calculate the water particle velocity and the pressure at a point 4 m ahead of the wave crest and 2 m below the stillwater level. Calculate the horizontal and vertical dimensions of the water particle orbit at this point.

c. Calculate the equivalent deep water height, length, celerity and energy.

· 2. a. What is the maximum height wave having a 3-sec period that can be generated in the tank of Prob. 1 with a water depth of 5 m?

b. What is the wave height and water depth at breaking for the 1-m high, 4-sec period wave of Prob. 1 when shoaling on a 1:10 slope in the tank? What type breaker would you expect?

· 3. A pressure gage located 1 m off the bottom in 10 m depth of water measures an average maximum pressure of 10 N/cm^2 having an average period of 12 sec. Find the wave height and length.

4. a. Derive an equation for the horizontal component of particle convective acceleration in a wave.

b. Calculate and compare the horizontal components of convective and local acceleration for a particle under the conditions of Prob. 1.b.

5. Demonstrate, using Eq. 2.41, that $C_g = C/2$ in deep water and $C_g = C$ in shallow water.

6. Derive the equations for the horizontal and vertical components of water particle velocity in a pure standing wave.

· 7. A wave with a 10-sec period and 2-m deep water height shoals without refracting on the smooth slope shown. Calculate the wave energy per m of crest length as the wave is about to break and calculate the wave runup above the swl.

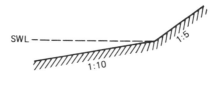

8. Sketch the pattern of streamlines in one length (crest to crest) of a progressive wave. Place arrows on this sketch to show the direction of fluid particle acceleration throughout the wave.

9. Consider a deep water wave having a height of 2 m and a period of 9 sec shoaling on a 1:50 slope without refraction. Calculate, for comparison,

the crest particle velocity in deep water, at $d = 20$ m, and just prior to breaking. Calculate the wave celerity just prior to breaking and compare it to the crest particle velocity. Comment on any discrepancies.

10. Calculate the energy head for the wave in Prob. 9 just before breaking and compare it to the wave runup (assume a smooth slope) if the slope landward of the stillwater level is 1:7.

11. A 10-sec. wave approaches shore. Calculate its length in water depths of 100 m, 50 m, 20 m and 5 m.

12. A 12-sec, 2-m high wave in deep water shoals without refracting. Calculate the maximum horizontal velocity component and the displacement from the mean position for a particle 5 m below the stillwater level in deep water and where the water depth is 8 m.

13. Equation 2.45 defines the surface position as a function of time for a standing wave. Calculate the potential energy per wave length and the potential energy density as a function of time. Realizing that at the instant a standing wave has zero velocity throughout, all energy is potential energy, determine the total and kinetic energies per wave length and the total and kinetic energy densities.

14. To an observer moving in the direction of a monochromatic wave train at the wave celerity, the wave motion appears steady. The surface particle velocity at the crest $U_c = C_0 + \pi H / T$ and at the trough $U_t = C_0 - \pi H / T$ for a deep water wave. Apply the Bernoulli equation to the surface to derive Eq. 2.16.

15. For a given wave height, period, and water depth, which of the following parameters depend on the water density: celerity, length, energy density, and particle pressure and velocity at a given depth? Explain each answer.

· 16. Waves with a period of 10 sec and a deep water height of 1 m arrive normal to shore. A 100-m long device that converts wave motion to electrical power is installed parallel to shore in water 6 m deep. If the device is 50 percent efficient what horsepower is produced?

17. Demonstrate that the velocity potential defined by Eq. 2.10 does represent irrotational flow.

Three

WAVE REFRACTION, DIFFRACTION, AND REFLECTION

As a train of waves travels toward the shore the wave crest pattern is usually modified by refraction, diffraction, and/or reflection.

Refraction occurs in transitional and shallow water where wave celerity decreases with decreasing water depth to cause the portion of the wave crest that is in shallower water to propagate forward at a slower speed than the portion in deeper water. The result is a bending of the wave crests so they approach the orientation of the bottom contours (see Fig. 3.1). Wave orthogonals will also bend and their convergence or divergence will lead to local increases or decreases in wave energy and, consequently, height.

Diffraction occurs when the height of a wave is greater at one point along the wave crest than at an adjacent point, causing a flow of energy along the crest in the direction of decreasing height and an adjustment of heights along the wave crest. This is important, for example, in causing the propagation of wave energy into the shadow region behind coastal structures that interrupt a train of waves.

Wave refraction, diffraction, and reflection along with shoaling effects will determine the nearshore wave height and crest orientation at a given location. Thus a spectrum of waves will be modified due to the relative increase or decrease in height of the various component waves. Quantification of these effects is important, for example, in predicting the longshore component of wave energy flux and thus the longshore transport of sediment. This information aids in establishing the best location for a harbor, for the harbor entrance once the harbor is located, and in determining the level of wave activity a particular coastal structure must withstand.

In this chapter we consider separately the effects of refraction, diffraction, and reflection on a train of long-crested monochromatic waves. This approach is only an approximation of reality as these effects occur simultaneously and are somewhat interdependent. In addition, real waves typically have a crest length only a few times greater than the wave length (Ralls and Wiegel, 1956). However, given the deep water wave period, height, and direction and the nearshore hydrography, one can predict the nearshore wave height and direction with sufficient accuracy for most engineering purposes. The components of a spectrum of waves can be analyzed separately and regrouped to approximate the nearshore characteristics of the wave spectrum. An easier but less desirable approach is to select a single design wave height and period to represent the spectrum of waves and to analyze the behavior of this wave as it approaches the shore.

3.1. WAVE REFRACTION

Figure 3.1 shows a hypothetical shoreline and nearshore bottom contours. A wave train with a deep water wave length, L_0, is approaching the shore with a crest orientation in deep water that is parallel to the average shoreline

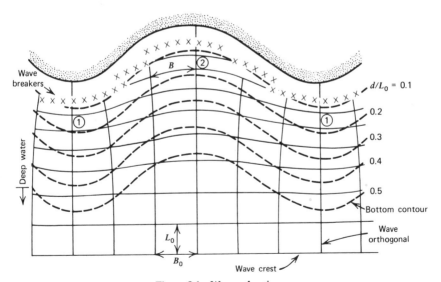

Figure 3.1 Wave refraction.

location. Bottom contour depths are given relative to the deep water wave length. As portions of the wave crest enter the region where $d/L_0 < 0.5$ the wave length and celerity decrease as given by Eq. 2.20. The result is a refraction of the wave train, with the wave crest orientations approaching the orientation of the bottom contours as shown. If one constructs equally spaced orthogonal lines along the deep water wave crests and extends these toward shore, being sure they are always normal to the wave crests, one can see the pattern of energy distribution at any point along a wave crest. Where orthogonals converge the energy per unit crest length increases and vice versa.

The convergence and divergence of wave orthogonals along with the effects of shoaling cause the wave height to vary according to (see Eq. 2.40)

$$\frac{H}{H_0} = \sqrt{\frac{L_0}{2nL}} \ \sqrt{\frac{B_0}{B}} = \frac{H}{H_0'} \sqrt{\frac{B_0}{B}} \qquad (3.1)$$

H/H_0', the shoaling coefficient, depends only on d/L or d/L_0 as shown in Fig. 2.3. The orthogonal spacing ratio B_0/B for the point of interest is determined from the refraction diagram as shown in Fig. 3.1. Note in Fig. 3.1 that refraction causes a convergence of orthogonals over the submerged ridge (point 1), resulting in higher waves that break further offshore. Over the trough (point 2), wave heights are lower than those over the ridge and can actually be much lower than the deep water height if the refraction effects overcome the wave height increase due to shoaling.

Since wave celerity depends on the wave period, waves with different periods will refract differently as they approach the shore. Longer period waves begin to feel bottom and refract in deeper water and consequently undergo greater refraction by the time they reach shore. For engineering purposes, refraction diagrams must be drawn for a number of wave periods to cover the range of important periods expected at a particular location. Also, a series of diagrams must be constructed to cover the range of deep water directions from which each period wave may arrive. From all of these refraction diagrams one can construct a plot similar to Fig. 3.2, which gives the refraction coefficient $K_R = \sqrt{B_0/B}$ at a shoreline site for the range of wave periods and directions expected. Given deep water wave heights one can determine the most critical wave heights to be expected at the site by applying Fig. 3.2 and Eq. 3.1. Interpolation from the refraction diagrams would give the crest orientation at the site for the critical waves. The modification of a spectrum of waves coming from a given direction can also be deduced from Fig. 3.2 and Eq. 3.1 by calculating the ratio of nearshore to deep water energy density (i.e. H^2/H_0^2) for each wave period.

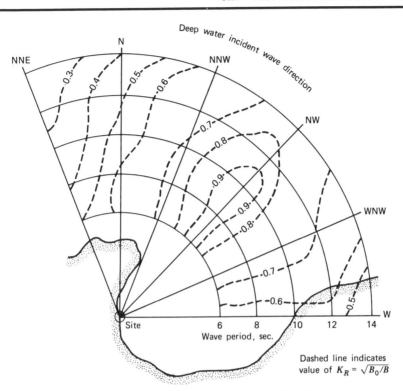

Figure 3.2   Wave refraction analysis for shoreline site.

3.2.   CONSTRUCTION OF WAVE REFRACTION DIAGRAMS

The first method developed for the construction of wave refraction diagrams is known as the wave crest method (Johnson et al., 1948). Beginning with a given wave crest position in deep water, points on the crest are advanced normal to the crest by an integral number of wave lengths, and the new crest position is drawn. This process is repeated until the refraction diagram is completed. Given the deep water wave length and the local average depth, the local wave length and thus the length of an integral number of wave lengths can be calculated from Eq. 2.20. A template that graphically solves Eq. 2.20 (see Wiegel, 1964) may be constructed to simplify the plotting of crest positions. After all the crest positions for a shoaling wave are drawn, orthogonal lines are constructed normal to the wave crest at desired intervals.

A second graphical method for constructing refraction diagrams, known as the orthogonal method (Arthur et al., 1952) is based upon Snell's Law,

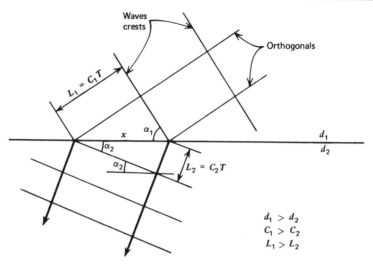

Figure 3.3 Definition sketch for Snell's Law derivation.

which may be derived by considering Fig. 3.3. A train of waves travels over a step (ignore wave reflection by the step) where the depth instantaneously decreases from d_1 to d_2, causing the wave celerity and length to decrease from C_1 and L_1 to C_2 and L_2 respectively. For an orthogonal spacing x and time interval T, $\sin\alpha_1 = C_1 T/x$ and $\sin\alpha_2 = C_2 T/x$. Dividing

$$\frac{\sin\alpha_1}{\sin\alpha_2} = \frac{C_1}{C_2} = \frac{L_1}{L_2} \tag{3.2}$$

which is Snell's Law. Applying Eq. 3.2 to wave refraction over a gradually varied slope, α_1 and α_2 become the angles between the wave crest and bottom contour at successive points along an orthogonal, and C_1 and C_2 become the wave celerities at the points where α_1 and α_2 are measured.

When waves shoal over nearshore contours that are essentially straight and parallel as shown in Fig. 3.4

$$\frac{\sin\alpha_0}{L_0} = \frac{\sin\alpha_1}{L_1} = x$$

if we choose B_0 and B_1 so the orthogonal lengths equal L_0 and L_1 as shown. Then

$$\frac{B_0}{\text{Cos}\,\alpha_0} = x = \frac{B_1}{\text{Cos}\,\alpha_1}$$

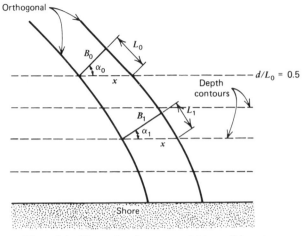

Figure 3.4 Wave refraction over straight parallel contours.

or

$$K_R = \sqrt{\frac{B_0}{B_1}} = \sqrt{\frac{\cos \alpha_0}{\cos \alpha_1}} \qquad (3.3)$$

where

$$\alpha_1 = \sin^{-1} \left(\frac{C_1}{C_0} \sin \alpha_0 \right) \qquad (3.4)$$

Equations 3.3 and 3.4 allow one to estimate the effects of wave refraction at a shoreline with uniform nearshore hydrography (see Probs. 1, 2). Wiegel and Arnold (1957) conducted wave tank tests of wave refraction on uniform slopes and found Snell's Law valid for deep water incidence angles between 10° and 70° and slopes ranging from 1:10 to vertical (step).

In order to develop a graphical technique for the orthogonal method, consider Fig. 3.5. A midcontour is drawn between existing bottom contours 1 and 2 and the incoming wave orthogonal is extended to this midcontour. Construct a line tangent to the midcontour at P, a line normal to the orthogonal from P to some arbitrary point R, and a line normal to the tangent line at Q to R. Then $\overline{RQ} = \overline{PR} \sin \alpha_1$. Construct $\overline{RS} = (C_1/C_2)\overline{PR}$ where C_1 and C_2 are the wave celerities at contours 1 and 2 respectively. Then

$$\sin \alpha_2 = \frac{\overline{RQ}}{\overline{RS}} = \frac{\overline{PR} \sin \alpha_1}{(C_1/C_2)\overline{PR}} = \frac{C_2}{C_1} \sin \alpha_1$$

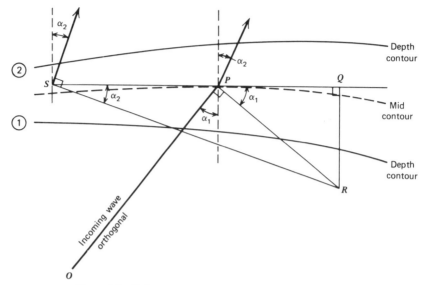

Figure 3.5 Definition sketch for orthogonal refraction method.

From Snell's law [$\sin \alpha_2 = (C_2/C_1)\sin \alpha_1$]; α_2 is the angle of departure of the orthogonal from the midcontour. The abrupt deflection of the orthogonal would occur if there was a step at the midcontour from d_1 to d_2. Usually, a uniform depth decrease from d_1 to d_2 is assumed and a curved orthogonal is drawn tangent to the constructed orthogonal at its intersections with contours 1 and 2.

Figure 3.6, developed from Fig. 3.5, is the template pattern for constructing refraction diagrams by the orthogonal method. The point R in Fig. 3.5 is represented by the turning point in Fig. 3.6, and P is the intersection of the three lines. A segment of a circle having a center at the turning point is constructed tangent to the orthogonal line, and marked in degrees $\Delta \alpha$. With the template constructed on transparent paper at a scale such that the distance from the turning point to the orthogonal is about 15 cm, one can construct refraction diagrams on most hydrographic charts.

The procedure for constructing refraction diagrams is as follows:

1. Locate the depth contour represented by $d/L_0 = 0.5$ on a hydrographic chart of the area of interest. Then label each of the shallower contours in terms of the relative depth d/L_0. Bottom contour irregularities that are smaller than the wave length do not appreciably affect the wave motion and may be smoothed out.

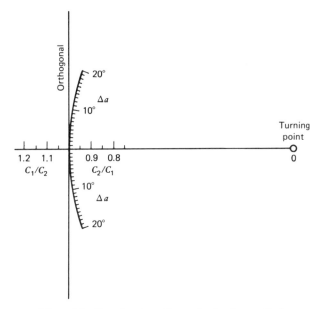

Figure 3.6 Template for orthogonal refraction method.

2. For each contour and the one landward of it calculate the ratio of wave celerities C_1/C_2, where C_1 is the celerity at the deeper contour. From Eq. 2.13

$$\frac{C_1}{C_2} = \frac{\tanh(2\pi d_1/L_1)}{\tanh(2\pi d_2/L_2)}$$

where d/L is a function of d/L_0 as given by Eq. 2.20 or Fig. 2.3.

3. Starting at the two most seaward contours, construct a midcontour, extend the incoming orthogonal to the midcontour, and construct a line tangent to the midcontour at the intersection of the midcontour with the orthogonal.

4. Lay the template (Fig. 3.6) with the line marked orthogonal over the incoming orthogonal and $C_1/C_2 = 1.0$ at the intersection of the midcontour and orthogonal.

5. Rotate the template about the turning point until the value of C_1/C_2 intersects the tangent to the midcontour. The line labelled orthogonal now lies in the direction of the departing orthogonal. However, it is not at the correct position of the departing orthogonal.

6. With a pair of triangles, move the departing orthogonal to a parallel position such that the lengths between contours of the incoming and departing orthogonals are equal.

7. Repeat the above procedure for successive contour intervals.

Orthogonals may be constructed from deep to shallow water using the same procedure, except the C_2/C_1 values are used where C_1 is still the wave celerity at the deeper contour.

To obtain the desired accuracy, Arthur et al. (1952) recommend a contour spacing such that $|\Delta C/C_1| < 0.2$ and $|\Delta \alpha| < 15°$. If the angle between the wave crest and bottom contour exceeds 80°, the outlined procedure is not sufficiently accurate and a modified procedure (see U.S. Army Coastal Engineering Research Center, 1973) should be used.

The orthogonal method is preferable to the wave crest method for general use. The wave crest method was briefly presented because it helps one to understand the phenomena of wave refraction. Of the two methods, the orthogonal method is more likely to be accurate (Dunham, 1950; Pierson et al., 1951). Also, it directly yields the orthogonal lines necessary for the calculation of wave height, energy density, and direction of wave travel. As wave crests are normal to wave orthogonals, wave crests can be added to the refraction diagram as desired.

A thorough wave refraction analysis for a coastal site requires the construction of many refraction diagrams, one for each combination of wave period and deep water direction of interest. This has led to the development of computer techniques for the numerical calculation of wave refraction diagrams. Once the computer program has been developed and offshore hydrographic data has been entered into storage, refraction diagrams can be constructed and automatically plotted much more rapidly and with less possibility of error than by manual techniques.

The equations presented by Munk and Arthur (1952) for the path of a wave orthogonal and for orthogonal spacing, both as a function of wave celerity (or wave period and water depth), are used as the basis for the numerical calculations. The specific approaches used by different authors (Griswold, 1963; Wilson, 1966; Dobson, 1967; and Jen, 1969) vary as to the solution techniques for the basic equations and to the interpolation schemes used to determine the depth at a desired point from the input grid of water depths. All depth interpolation schemes (e.g. method of least squares) involve a degree of smoothing and thus affect to some extent the results obtained.

Some procedures just calculate orthogonal positions from which wave heights can be calculated manually; others calculate and plot wave height values as well as orthogonal positions. Skovgaard et al. (1975) included bottom friction in their numerical wave refraction method and Smith and Camfield (1972) included the breaking and reforming of wave trains in shoaling waters and over submerged reefs.

3.3. WAVE REFRACTION STUDY

As part of a project to forecast the wave climate for the New Jersey coast, Pierson (1950) and Pierson et al. (1951) conducted a detailed wave refraction study for the offshore area near Long Branch, New Jersey (Fig. 3.7a).

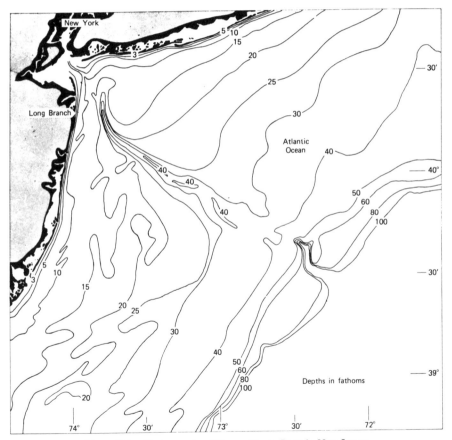

Figure 3.7a Hydrography offshore of Long Branch, New Jersey.

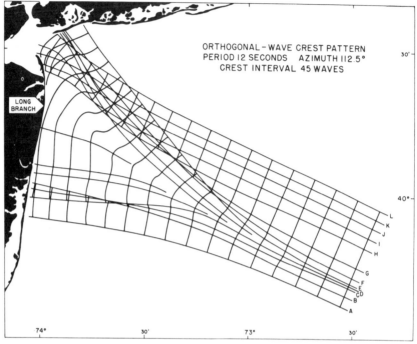

Figure 3.7b   Refraction diagram, 12=sec wave from 112.5° azimuth (Pierson, 1950).

The manual orthogonal method was used to construct refraction diagrams for a range of wave periods and deep water directions representative of the area. After the orthogonal patterns were constructed, wave crests were drawn by calculating the wave advance along each orthogonal for a given number of wave lengths and connecting these points to form wave crests. In order to obtain worthwhile results, bottom contour patterns for the very irregular offshore hydrography were considerably smoothed; the degree of smoothing depending in an arbitrary way on the map scale and wave period.

Figure 3.7b is the refraction diagram for a 12-sec wave approaching shore from a deepwater azimuth of 112.5°. The submarine canyon has a rather drastic effect on the wave orthogonal pattern. Orthogonals E and F cross to cause an overlapping of wave orthogonals and the development of a pair of wave trains approaching the shore at slightly different directions. The wave energy between orthogonals E and L is concentrated near the entrance to New York Harbor. Overlapping of wave orthogonals and the development of additional wave crests also occurs between orthogonals B and D. Wave diffraction occurs at the ends of broken wave crests along orthogonals B, D, E, H, and so on.

The wave energy between D and E is spread over a long section of coastline. As a result, the nearshore wave energy at Long Branch is extremely low; the wave refraction coefficient, K_R, being approximately 0.17. However, for the same offshore wave direction but a 6-sec wave period the value of K_R is 0.9. Also, 12-sec waves having an offshore azimuth of 90° result in a K_R of 0.75. This demonstrates the possible importance of wave refraction on the wave climate in a particular area.

Figure 3.7*b* also demonstrates the difficulty one can have in deciding where to start wave orthogonals in deep water in order to analyze conditions at a particular shore location. The only approach is to continue to construct additional orthogonals until the desired result is achieved. When the refraction coefficient is determined from orthogonal spacing, it represents the average value, for the region between orthogonals and local variations can be large, as demonstrated by Fig. 3.7*b*. Care must be taken to use as small a spacing as possible within the limits of the accuracy of the refraction diagram construction method used.

3.4. REFRACTION BY CURRENTS

When waves travel from a still water area to an area where a current exists, the wave period remains constant, the wave height, length, and celerity change, and, if the wave orthogonals are not parallel to the current direction, the waves refract. This is of practical importance, for example, when waves approach a river mouth or tidal entrance.

Consider Fig. 3.8, where waves are moving from a still water area (a) to a channel (b) where the current velocity is normal to the incident wave crests and has a magnitude V. As there can be no steady accumulation or disappearance (assume no wave breaking) of waves between a and b, the same number of waves pass a and b in a given period of time. Thus the wave period must be the same at a and b. With C_b equal to the wave celerity relative to the water velocity at b

$$\frac{L_a}{C_a} = T = \frac{L_b}{V + C_b} \qquad (3.5)$$

Assuming deep water, from Eq. 2.15

$$\frac{L_b}{L_a} = \frac{C_b^2}{C_a^2} \qquad (3.6)$$

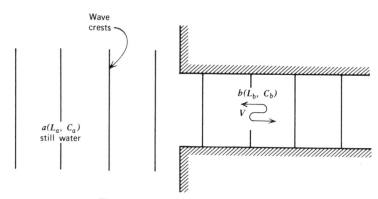

Figure 3.8   Waves encountering a current.

Combining these two equations and solving,

$$\frac{C_b}{C_a} = \frac{1}{2}\left[1 + \sqrt{1 + \frac{4V}{C_a}}\,\right] \tag{3.7}$$

When the current opposes the waves, V/C_a is negative so $C_b < C_a$ and $L_b < L_a$. From a conservation of wave power (Eq. 2.40) the wave height and thus the wave steepness must increase. The reverse is true when the waves and current move in the same direction.

If an opposing current causes a sufficient increase in wave steepness the waves will become unstable and break. Creation of an opposing current by a surface water jet or rising air bubbles has been used to dissipate wave energy, primarily by wave breaking.

Figure 3.9 depicts waves obliquely approaching a current from still water. As in the derivation of Snell's Law

$$\frac{L_a}{L_b} = \frac{\sin \alpha_a}{\sin \alpha_b} \tag{3.8}$$

Since the wave crest is continuous across the discontinuity,

$$V + \frac{C_b}{\sin \alpha_b} = \frac{C_a}{\sin \alpha_a} \tag{3.9}$$

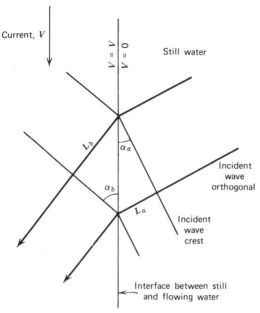

Figure 3.9 Wave refraction by a current.

Assuming deep water and solving Eqs. 3.6, 3.8, and 3.9

$$\frac{L_a}{L_b} = \left(1 - \frac{V}{C_a}\sin\alpha_a\right)^2 \tag{3.10}$$

See Johnson (1947) for details of the solution.

A current acting on a wave train, as shown in Fig. 3.9, causes an increase in wave steepness in two ways. First, the wave orthogonals come closer together, and second, the current acts on the wave length in the same fashion as in Fig. 3.8. The conservation of power equation, including the effects of refraction (Eq. 3.1), can be used to calculate changes in wave height.

3.5. WAVE DIFFRACTION

When a train of waves passes an impermeable structure there will be a transfer of wave energy along the wave crest into the lee of the structure as demonstrated by Fig. 3.10. As a result, the wave height in the region inside

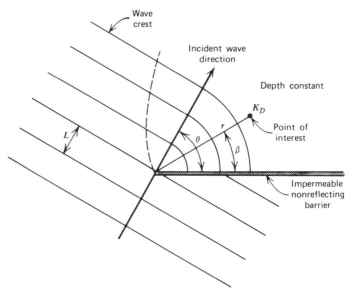

Figure 3.10   Wave diffraction behind a barrier.

the dashed line will be affected. The ratio of the wave height at a point in that region to the incident wave height is known as the diffraction coefficient K_D, where $K_D = f(\theta, \beta, r/L)$. Figure 3.10 defines θ, β, and r. The wave crest pattern in the lee of the structure can be approximated by concentric circular arcs. Note that the water depth in Fig. 3.10 is constant; otherwise, the wave crest pattern and heights would also be affected by refraction. Generally, the wave height decreases along the wave crest into the lee of the structure.

If the barrier in Fig. 3.10 reflected wave energy, the reflected wave crests (see Section 3.7) would also diffract to form concentric circular wave crests around the tip of the barrier.

Water wave diffraction is analogous to the diffraction of light. Penny and Price (1952) showed that the mathematical solution for the diffraction of light can also be used to predict the wave crest pattern and height variation for diffracted water waves. A summary of the solution is presented by Wiegel (1964) and by Putnam and Arthur (1948) who also verified the solution by a model study. Johnson (1952, 1953) simplified the Penny and Price solution and presented the results in the form of diffraction coefficient contours on cartesian coordinate diagrams. Wiegel (1962) used the exact solution presented by Penny and Price to calculate K_D as a function of selected values of

θ, β, and r/L. His results are tabulated in Table 3.1.

Consider, as an example, a train of 6-sec waves approaching an impermeable, nonreflecting breakwater at an angle θ of 60°, as shown in Fig. 3.10. If the water depth is constant at 10 m, the wave length, from Eq. 2.14, is 48.3 m. At an angle of 30° and a distance of 96.6 m from the breakwater tip, K_D is 0.28 (i.e. $\alpha = 30°$, $r/L = 2.0$ in Table 3.1). Thus a 1 m high incident wave would be 0.28 m high at this point and moving in the direction of the radial line that defines α. This method for evaluating effects of wave diffraction can be applied to both ends of a detached offshore breakwater provided the length of the breakwater is equal to or greater than a few wave lengths (see Probs. 12 and 13).

For values of θ, β, and r/L intermediate between those given by Table 3.1, one must apply linear interpolation. Fan et al. (1967) wrote a computer program to solve the Penny and Price equations and presented the results for many times the number of values given by Table 3.1. This should be used if greater accuracy is desired.

At a particular point of interest in the lee of a structure, the diffraction coefficient depends on the incident wave direction and the wave length (or period). A spectrum of waves, if all are coming from the same direction, will generally experience a greater percentage decrease in energy density at successively lower wave periods. Thus the energy density concentration will shift toward the higher wave periods of the spectrum. By determining K_D for the range of incident wave periods and directions one can evaluate the modified characteristics for a spectrum of waves at a point of interest in the lee of a structure.

The theory for diffraction of water waves approaching in a direction normal to a long straight structure and passing through a single gap in that structure was also developed by Penny and Price (1952). Their theoretical work was in agreement with the results of laboratory studies conducted by Blue and Johnson (1949). Johnson (1952, 1953) presented plots, similar to Fig. 3.11, which give K_D contours in the lee of a gap for different gap width to incident wave length ratios. As in Fig. 3.10, horizontal scales are made nondimensional by dividing by the incident wave length. Beyond two or three wave lengths from the gap, wave crests are essentially circular arcs concentric about the midpoint of the gap.

Johnson demonstrated that these diagrams could be used if the angle of wave incidence is other than 90° by using a projected imaginary gap width as shown in Fig. 3.12. When the gap width is about five times the incident wave length or greater, the diffraction effects of the barrier on each side of the gap are essentially independent. Then the theory for diffraction at the end of a single barrier can be used for each side.

TABLE 3.1. WAVE DIFFRACTION COEFFICIENTS, K_D, AS A FUNCTION (WIEGEL, 1962)

r/L	0	15	30	45	60	75	90	105	120	135	150	165	180
						$\theta = 15°$							
1/2	0.49	0.79	0.83	0.90	0.97	1.01	1.03	1.02	1.01	0.99	0.99	1.00	1.00
1	0.38	0.73	0.83	0.95	1.04	1.04	0.99	0.98	1.01	1.01	1.00	1.00	1.00
2	0.21	0.68	0.86	1.05	1.03	0.97	1.02	0.99	1.00	1.00	1.00	1.00	1.00
5	0.13	0.63	0.99	1.04	1.03	1.02	0.99	0.99	1.00	1.01	1.00	1.00	1.00
10	0.35	0.58	1.10	1.05	0.98	0.99	1.01	1.00	1.00	1.00	1.00	1.00	1.00
						$\theta = 30°$							
1/2	0.61	0.63	0.68	0.76	0.87	0.97	1.03	1.05	1.03	1.01	0.99	0.95	1.00
1	0.50	0.53	0.63	0.78	0.95	1.06	1.05	0.98	0.98	1.01	1.01	0.97	1.00
2	0.40	0.44	0.59	0.84	1.07	1.03	0.96	1.02	0.98	1.01	0.99	0.95	1.00
5	0.27	0.32	0.55	1.00	1.04	1.04	1.02	0.99	0.99	1.00	1.01	0.97	1.00
10	0.20	0.24	0.54	1.12	1.06	0.97	0.99	1.01	1.00	1.00	1.00	0.98	1.00
						$\theta = 45°$							
1/2	0.49	0.50	0.55	0.63	0.73	0.85	0.96	1.04	1.06	1.04	1.00	0.99	1.00
1	0.38	0.40	0.47	0.59	0.76	0.95	1.07	1.06	0.98	0.97	1.01	1.01	1.00
2	0.29	0.31	0.39	0.56	0.83	1.08	1.04	0.96	1.03	0.98	1.01	1.00	1.00
5	0.18	0.20	0.29	0.54	1.01	1.04	1.05	1.03	1.00	0.99	1.01	1.00	1.00
10	0.13	0.15	0.22	0.53	1.13	1.07	0.96	0.98	1.02	0.99	1.00	1.00	1.00
						$\theta = 60°$							
1/2	0.40	0.41	0.45	0.52	0.60	0.72	0.85	1.13	1.04	1.06	1.03	1.01	1.00
1	0.31	0.32	0.36	0.44	0.57	0.75	0.96	1.08	1.06	0.98	0.98	1.01	1.00
2	0.22	0.23	0.28	0.37	0.55	0.83	1.08	1.04	0.96	1.03	0.98	1.01	1.00
5	0.14	0.15	0.18	0.28	0.53	1.01	1.04	1.05	1.03	0.99	0.99	1.00	1.00
10	0.10	0.11	0.13	0.21	0.52	1.14	1.07	0.96	0.98	1.01	1.00	1.00	1.00
						$\theta = 75°$							
1/2	0.34	0.35	0.38	0.42	0.50	0.59	0.71	0.85	0.97	1.04	1.05	1.02	1.00
1	0.25	0.26	0.29	0.34	0.43	0.56	0.75	0.95	1.02	1.06	0.98	0.98	1.00
2	0.18	0.19	0.22	0.26	0.36	0.54	0.83	1.09	1.04	0.96	1.03	0.99	1.00
5	0.12	0.12	0.13	0.17	0.27	0.52	1.01	1.04	1.05	1.03	0.99	0.99	1.00
10	0.08	0.08	0.10	0.13	0.20	0.52	1.14	1.07	0.96	0.98	1.01	1.00	1.00
						$\theta = 90°$							
1/2	0.31	0.31	0.33	0.36	0.41	0.49	0.59	0.71	0.85	0.96	1.03	1.03	1.00
1	0.22	0.23	0.24	0.28	0.33	0.42	0.56	0.75	0.96	1.07	1.05	0.99	1.00
2	0.16	0.16	0.18	0.20	0.26	0.35	0.54	0.69	1.08	1.04	0.96	1.02	1.00
5	0.10	0.10	0.11	0.13	0.16	0.27	0.53	1.01	1.04	1.05	1.02	0.99	1.00
10	0.07	0.07	0.08	0.09	0.13	0.20	0.52	1.14	1.07	0.96	0.99	1.01	1.00

54

r/L	β (Degrees)												
	0	15	30	45	60	75	90	105	120	135	150	165	180
					$\theta = 105°$								
1/2	0.28	0.28	0.29	0.32	0.35	0.41	0.49	0.59	0.72	0.85	0.97	1.01	1.00
1	0.20	0.20	0.24	0.23	0.27	0.33	0.42	0.56	0.75	0.95	1.06	1.04	1.00
2	0.14	0.14	0.13	0.17	0.20	0.25	0.35	0.54	0.83	1.08	1.03	0.97	1.00
5	0.09	0.09	0.10	0.11	0.13	0.17	0.27	0.52	1.02	1.04	1.04	1.02	1.00
10	0.07	0.06	0.08	0.08	0.09	0.12	0.20	0.52	1.14	1.07	0.97	0.99	1.00
					$\theta = 120°$								
1/2	0.25	0.26	0.27	0.28	0.31	0.35	0.41	0.50	0.60	0.73	0.87	0.97	1.00
1	0.18	0.19	0.19	0.21	0.23	0.27	0.33	0.43	0.57	0.76	0.95	1.04	1.00
2	0.13	0.13	0.14	0.14	0.17	0.20	0.26	0.16	0.55	0.83	1.07	1.03	1.00
5	0.08	0.08	0.08	0.09	0.11	0.13	0.16	0.27	0.53	1.01	1.04	1.03	1.00
10	0.06	0.06	0.06	0.07	0.07	0.09	0.13	0.20	0.52	1.13	1.06	0.98	1.00
					$\theta = 135°$								
1/2	0.24	0.24	0.25	0.26	0.28	0.32	0.36	0.42	0.52	0.63	0.76	0.90	1.00
1	0.18	0.17	0.18	0.19	0.21	0.23	0.28	0.34	0.44	0.59	0.78	0.95	1.00
2	0.12	0.12	0.13	0.14	0.14	0.17	0.20	0.26	0.37	0.56	0.84	1.05	1.00
5	0.08	0.07	0.08	0.08	0.09	0.11	0.13	0.17	0.28	0.54	1.00	1.04	1.00
10	0.05	0.06	0.06	0.06	0.07	0.08	0.09	0.13	0.21	0.53	1.12	1.05	1.00
					$\theta = 150°$								
1/2	0.23	0.23	0.24	0.25	0.27	0.29	0.33	0.38	0.45	0.55	0.68	0.83	1.00
1	0.16	0.17	0.17	0.18	0.19	0.22	0.24	0.29	0.36	0.47	0.63	0.83	1.00
2	0.12	0.12	0.12	0.13	0.14	0.15	0.18	0.22	0.28	0.39	0.59	0.86	1.00
5	0.07	0.07	0.08	0.08	0.08	0.10	0.11	0.13	0.18	0.29	0.55	0.99	1.00
10	0.05	0.05	0.05	0.06	0.06	0.07	0.08	0.10	0.13	0.22	0.54	1.10	1.00
					$\theta = 165°$								
1/2	0.23	0.23	0.23	0.24	0.26	0.28	0.31	0.35	0.41	0.50	0.63	0.79	1.00
1	0.16	0.16	0.17	0.17	0.19	0.20	0.23	0.26	0.32	0.40	0.53	0.73	1.00
2	0.11	0.11	0.12	0.12	0.13	0.14	0.16	0.19	0.23	0.31	0.44	0.68	1.00
5	0.07	0.07	0.07	0.07	0.08	0.09	0.10	0.12	0.15	0.20	0.32	0.63	1.00
10	0.05	0.05	0.05	0.06	0.06	0.06	0.07	0.08	0.11	0.11	0.21	0.58	1.00
					$\theta = 180°$								
1/2	0.20	0.25	0.23	0.24	0.25	0.28	0.31	0.34	0.40	0.49	0.61	0.78	1.00
1	0.10	0.17	0.16	0.18	0.18	0.23	0.22	0.25	0.31	0.38	0.50	0.70	1.00
2	0.02	0.09	0.12	0.12	0.13	0.18	0.16	0.18	0.22	0.29	0.40	0.60	1.00
5	0.02	0.06	0.07	0.07	0.07	0.08	0.10	0.12	0.14	0.18	0.27	0.46	1.00
10	0.01	0.05	0.05	0.04	0.06	0.07	0.07	0.08	0.10	0.13	0.20	0.36	1.00

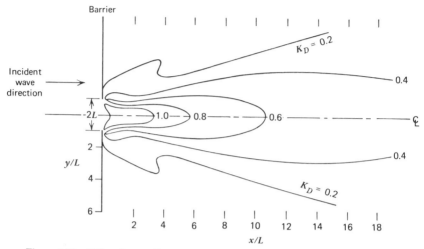

Figure 3.11   Diffraction coefficients at barrier gap; gap width $= 2L$ (Johnson, 1952).

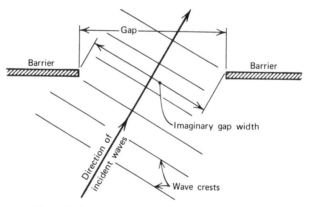

Figure 3.12   Oblique wave incident to a barrier gap.

Carr and Stelzriede (1952) applied the Morse and Rubenstein (1938) theory of wave diffraction by slits to water waves passing through a barrier gap. They also performed limited experiments demonstrating the validity of the theory. Johnson (1952) used Carr and Stelzriede's results to present a series of K_D contour diagrams for a range of incident wave angles and a gap opening width equal to one wave length. A compilation of most of the K_D contour diagrams for single barriers and barrier gaps discussed in this section is presented by the U.S. Army Coastal Engineering Research Center (1973).

3.6. COMBINED REFRACTION AND DIFFRACTION

In most instances when waves are diffracting, the bottom hydrography is such as to cause simultaneous wave refraction. To date, a technique for predicting the combined effects of wave refraction and diffraction in a region with natural bottom contours has not been developed. The U.S. Army Corps of Engineers (1973) suggests the following approach for refracting waves interrupted by a barrier: (1) refract the incident wave shoreward to the barrier; (2) construct the diffraction pattern in the lee of the barrier for a distance of three or four wave lengths, if possible; and (3) with the wave crest and orthogonal orientation from the last wave crest in the diffracted pattern continue the refraction diagram to the point of interest. The ratio of the wave height at the point of interest to the deep water wave height will equal the product of: (1) the refraction coefficient at the barrier; (2) the diffraction coefficient at the last diffracted wave crest; (3) the refraction coefficient from the last diffracted wave crest to the point of interest; and (4) the shoaling coefficient from deep water to the point of interest.

The choice of the number of wave lengths to carry the diffraction analysis should fit the particular circumstances. As long as the bottom is approximately horizontal or the depth is greater than approximately 0.3 L_0, continue the diffraction analysis, as diffraction effects will very likely predominate. If the bottom is not horizontal or deeper than 0.3 L_0, but the wave crests approximately parallel the bottom contours, continue the diffraction analysis but include the effects of shoaling if you wish to evaluate the wave height in the diffraction zone.

Mobarek (1962) conducted laboratory experiments in a wave tank that had a single barrier oriented parallel to the wave generator and an opening from the barrier tip to the other wall, to represent half a breakwater gap. There was a 1:12 sloped shoaling bottom in the lee of the barrier with straight bottom contours normal to the barrier. For $d/L = 0.14$ at the gap the experimental wave heights on the slope generally agreed with the heights determined by the suggested procedure, which was applied by constructing a diffraction diagram for one wave length in the lee of the barrier and then conducting the refraction analysis to the point of interest on the slope.

Whenever the wave height is not constant along a wave crest, diffraction occurs. This is the case during most instances of wave refraction even though the wave crest may not be cut by some barrier. Diffractive effects can be neglected for most purely refraction analyses but, one should remember if there is an abrupt change in orthogonal spacing along a wave crest, diffraction can significantly affect the resulting wave heights. Diffraction will diminish the effect of wave refraction on wave height variation along a wave crest.

3.7. WAVE REFLECTION

Section 2.7 covers the two-dimensional aspects of wave reflection. Here, the reflected wave crest pattern and resulting wave heights for waves that obliquely approach a reflecting barrier will be considered. A graphical technique for determining the reflected wave crest pattern is demonstrated by Fig. 3.13. The incident wave crest is extended past the reflecting barrier with the orientation it would have if the imaginary region had a mirror image hydrography of the real hydrography. The mirror image of the imaginary incident wave crest extension is the reflected wave crest. If we define a reflection coefficient for the barrier C_R equal to the reflected wave height divided by the incident wave height, the amplitude of the reflected wave crest is the equivalent amplitude in the imaginary region times the reflection coefficient of the barrier. (See Fig. 3.13.)

Consider a wave train approaching a corner as shown in Fig. 3.14. The image and reflected crests for two waves have been constructed, but the diffraction that would occur at the ends of the reflected wave crests is not shown. The total vertical displacement of the water surface at any point is approximately the sum of the displacements of the incident and reflected waves at that point. As wave travel continues the reflected waves are reflected again and again and the pattern becomes complex and confused, particularly if the wall reflection coefficients are high. Carr (1952) and

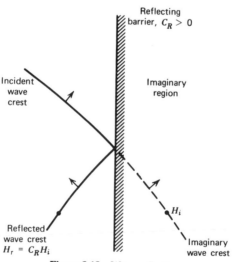

Figure 3.13   Wave reflection.

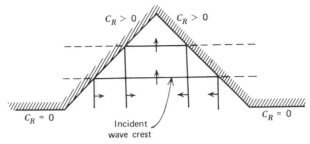

$c_R > 0$ $c_R > 0$

$c_R = 0$

Incident
wave crest

$c_R = 0$

Figure 3.14 Wave reflection at a corner, constant water depth.

Ippen (1966) discuss this in greater detail and present suggestions for applying these concepts to the study of wave reflection in harbors.

3.8. SUMMARY

Several aspects of the three-dimensional behavior of waves, particularly in the nearshore zone, have been presented in this chapter. The techniques presented are important in predicting nearshore wave characteristics, including height, energy, power, and direction of travel, which in turn are important to certain engineering problems such as evaluating wave characteristics in harbors and alongshore, determining wave forces on coastal structures, and investigating coastal sediment processes.

3.9. REFERENCES

Arthur, R. S., W. H. Munk and J. D. Isaacs (1952), "The Direct Construction of Wave Rays," *Transactions, American Geophysical Union*, Vol. 33, pp. 855–865.

Blue, F. L. and J. W. Johnson (1949), "Diffraction of Water Waves Passing Through a Breakwater Gap," *Transactions, American Geophysical Union*, Vol. 30, pp. 705–718.

Carr, J. H. (1952), "Wave Protection Aspects of Harbor Design," Rept. E-11, Hydromechanics Lab, Cal. Tech., 99 p.

Carr, J. H. and M. E. Stelzriede (1952), "Diffraction of Water Waves by Breakwaters," *Proceedings, Symposium on Gravity Waves*, National Bureau of Standards Circular 521, pp. 109–125.

Dobson, R. S. (1967), "Some Applications of a Digital Computer to Hydraulic Engineering Problems," Tech Rept. 80, Civil Engrg Dept., Stanford University.

Dunham, J. W. (1950), "Refraction and Diffraction Diagrams," *Proceedings, First Conference on Coastal Engineering*, Council on Wave Research, University of California, Berkeley, pp. 33–49.

Fan, S. S., J. E. Cumming and R. L. Wiegel (1967), "Computer Solution of Wave Diffraction by Semi-infinite Breakwater," HEL-1-8, University of California, Berkeley, 376 p.

Griswold, G. M. (1963), "Numerical Calculation of Wave Refraction," *Journal of Geophysical Research*, Vol. 68, pp. 1715–1723.

Ippen, A. T. (1966), *Estuary and Coastline Hydrodynamics*, McGraw-Hill, N.Y. pp. 318–323.

Jen, Y. (1969), "Wave Refraction Near San Pedro Bay, California," *Journal, Waterways and Harbors Division*, American Society of Civil Engineers, Vol. 95, pp. 379–393.

Johnson, J. W. (1947), "The Refraction of Surface Waves by Currents," *Transactions, American Geophysical Union*, Vol. 28, pp. 867–874.

Johnson, J. W., M. P. O'Brien and J. D. Isaacs (1948), "Graphical Construction of Wave Refraction Diagrams," Navy Hydrographic Office Publ. No. 605, Washington, D.C., 45 p.

Johnson, J. W. (1952), "Generalized Wave Diffraction Diagrams," *Proceedings, Second Conference on Coastal Engineering*, Council on Wave Research, Berkeley, pp. 6–23.

Johnson, J. W. (1953), "Engineering Aspects of Diffraction and Refraction," *Transactions, American Society of Civil Engineers*, Vol. 118, pp. 617–652.

Mobarek, I. (1962), "Effect of Bottom Slope on Wave Diffraction," HEL 1-1, University of California, Berkeley, 88 p.

Morse, P. M. and P. J. Rubinstein (1938), "The Diffraction of Waves by Ribbons and Slits," *Physical Reviews*, Vol. 54, pp. 895–898.

Penny, W. G. and A. T. Price (1952), "The Diffraction Theory of Sea Waves by Breakwaters and the Shelter Afforded by Breakwaters," *Philosophical Transactions, Royal Society*, Series A, Vol. 244, London, pp. 236–253.

Pierson, W. J. (1950), "The Interpretation of Crossed Orthogonals in Wave Refraction Phenomena," Tech. Memo 21, U.S. Army Beach Erosion Board, Washington, D.C., 83 p.

Pierson, W. J., D. P. Martineau, R. W. James and L. S. Pocinki (1951), "Ocean Wave Refraction Data for the Northern New Jersey Coast," Dept. of Meterology Report, New York University, 50 p.

Putnam, J. A. and R. S. Arthur (1948), "Diffraction of Water Waves by Breakwaters," *Transactions, American Geophysical Union*, Vol. 29, pp. 481–490.

Ralls, G. C. and R. L. Wiegel (1956), "A Laboratory Study of Short-crested Wind Waves," Tech. Memo 81, U.S. Army Beach Erosion Board, Washington, D.C., 28 p.

Skovgaard, O., I. G. Jonsson and J. A. Bertelsen (1975) "Computation of Wave Heights Due to Refraction and Friction," *Journal, Waterways, Harbors and Coastal Engineering Division*, American Society of Civil Engineers, Vol. 101, pp. 15–32.

Smith, B. S. L. and F. E. Camfield (1972), "A Refraction Study and Program for Periodic Waves Approaching a Shoreline, and Extending Beyond the Breaking Point," Tech. Report 16, Coll. of Marine Studies, Univ. of Delaware, 117 p.

U.S. Army Coastal Engineering Research Center (1973), *Shore Protection Manual*, 3 Vols., U.S. Government Printing Office, Washington, D.C.

Wiegel, R. L. (1957), "Model Study of Wave Refraction," Tech. Memo 103, U.S. Army Beach Erosion Board, Washington, D.C., 31 p.

Wiegel, R. L. (1962), "Diffraction of Waves by Semi-infinite Breakwater," *Journal, Hydraulics Division*, American Society of Civil Engineers, Vol. 88, pp. 27–44.

Wiegel, R. L. (1964), *Oceanographical Engineering*, Prentice-Hall, Englewood Cliffs, N.J., 532 p.

Wilson, W. S. (1966), "A Method for Calculating and Plotting Surface Wave Rays," Tech. Memo 17, U.S. Army Coastal Engineering Research Center, 57 p.

3.10. PROBLEMS

.1. A wave train is observed approaching a coast that has straight parallel near-shore contours oriented in the north-south direction. Where the depth is 5 m the wave length is 85 m and the wave crest forms an angle of 9° with the shore (waves from southwest). What is the incident wave direction in deep water?

·2. A wave train approaches the same shore location as in Prob. 1 but the deep water wave crests form an angle with the shore of 50°. If $T = 11$ seconds and $H_0 = 2$ m what is the wave height and the angle between the wave crest and shore where the depth is 6 m? At what depth will the wave break?

·3. A train of waves having a 1-sec period and a height of 5 cm is generated in a wave tank having the dimensions and water depths shown below. The side walls and step are lined with 100 percent effective wave absorbers. Construct the wave refraction diagram, indicate regions where wave diffraction effects would be significant, sketch the wave crest pattern in this region, and calculate the wave height at point A.

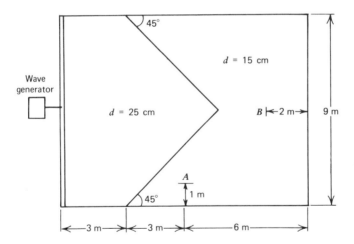

4. Given the hydrography in Prob. 5, construct a pair of orthogonals for an 11-sec wave approaching the contours at a 45° angle. If the wave is 3.5 m high in deep water, determine the wave height, water depth and angle between the wave crest and bottom contour just prior to wave breaking.

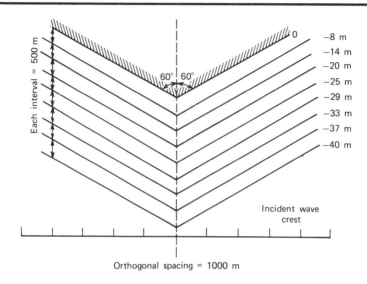

Orthogonal spacing = 1000 m

5. Draw eight parallel lines 3 cm apart and assign depths of 110, 100, 87, 73, 58, 40, 23, and 3 m. Starting at the same deep water point construct orthogonals for each of three waves approaching the contours at a 45° angle and having wave periods of 10, 11, and 12 sec, respectively.

6. A train of waves ($T = 7$ sec) approaches the shoreline shown above from the direction indicated. Construct the wave refraction diagram using the wave crest method. Draw wave orthogonals commencing at the points marked. Select one of these orthogonals and reconstruct it using the orthogonal method.

7. Waves having a period of 4 sec and a height of 1.5 m in deep water enter a harbor through a pair of parallel jetties aligned normal to the wave crests. What is the maximum ebb tidal velocity possible between the jetties without the waves breaking if the channel depth is 15 m? If the depth is 6 m?

8. Eight-sec waves approaching the east coast of the U.S. cross the Gulf Stream, which flows northeast. If the waves come from due east and the average Gulf Stream velocity is 1.5 m/sec, what is the direction of wave travel, the wave celerity, and percent change in wave height in the Gulf Stream?

9. Considering the effects of wave diffraction, what is the wave height at point B along the centerline in Prob. 3?

▲10. Consider the L-shaped breakwater that protects a harbor region dredged to a depth of 6 m as shown on the following page. For a 2 m high 6-sec period incident wave at the tip of the breakwater having the direction shown, what length x must the offshore arm have to diminish the wave height to 0.5 m at point A?

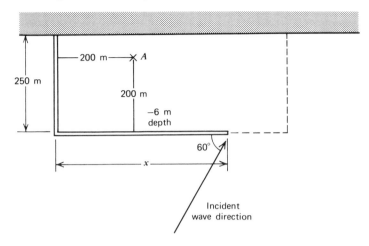

•11. With $x = 200$ m in Prob. 10 and the same incident wave direction, what are the values of K_D at point A for a 4-, 6-, 8-, 10-, and 12-sec wave? From this, discuss the effect of diffraction on a spectrum of waves all having the same incident direction.

12. Using NOS Chart 5386 (San Luis Obispo Bay and Approaches) determine the wave height at the tip of the Point San Luis breakwater and at the seaward end of the Point San Luis Wharf for a 9-sec, 4 m high deep water wave approaching the area from due west. At what depth along the wharf would this wave break?

13. A 7-sec, 2.5 m high deep water wave approaches shore in a normal direction. At a distance of 300 m from shore a 700 m long breakwater is constructed parallel to shore and the area leeward of the breakwater is dredged to −5 m, which is the natural water depth at the breakwater. What is the maximum wave height that will occur at the center of the breakwater on the leeward side?

14. If the deep water incident wave direction is 30° off a line normal to shore in Prob. 13 and the offshore bottom contours are all parallel to shore, what is the wave height at the center of the breakwater on the leeward side? Draw the wave crest pattern in the lee of the structure (ignore wave reflection).

•15. In Prob. 10, with the same incident wave height, period and deep water direction, what is the wave height at point A if the water depth in the lee of the breakwater decreases linearly from −6 m at the offshore arm to −1.5 m at point A?

16. A small harbor has a constant depth and the shape and reflection coefficients shown. For an incident wave having a height of 3 m, a length of 50 m and the direction shown, construct the wave crest pattern and determine the maximum wave height that will occur at point A.

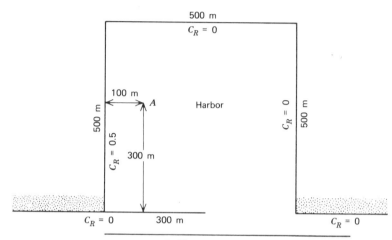

Incident wave crest

Four

COASTAL WATER
LEVEL FLUCTUATIONS

Chapters 2 and 3 are primarily concerned with waves having periods common to the wind-generated portion of the surface wave energy spectrum (Fig. 2.1). This chapter deals with certain longer period waves and other water level fluctuations that are important to coastal engineers. In particular, these may be classified as

1. Astronomical tide
2. Tsunamis
3. Basin oscillations
4. Storm surge
5. Climatologic effects
6. Geologic effects

Climatologic and geologic effects will be discussed briefly; major emphasis will be placed on the astronomical tide, tsunamis, basin oscillations, and storm surge.

The tide has primary periods around 12.4 and 24 hr while tsunami waves (siesmically generated sea surface waves) have periods that are typically in the range of 5–60 min. Basin oscillations are a resonant response to excitation by some segment of the sea's energy spectrum. The periods of resonant oscillation will depend on the geometry of the basin and on the periods of the available excitation energy. They may vary from a few minutes for a small harbor to a few hours for a large bay. Storm surge is a rise and fall of sea level caused primarily by wind stress and atmospheric pressure variations having a typical duration from several hours to a few days. Climatologic effects usually have seasonal or much longer periods as do most geologic effects.

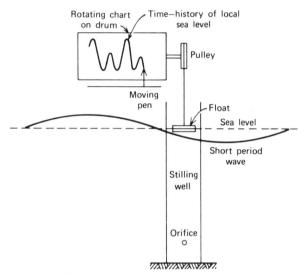

Figure 4.1   Float-stilling well tide gage.

Figure 4.1 is a schematic depiction of a common type of gage that can be used to measure the water level fluctuations discussed in this chapter. Sea level fluctuations cause the water level and float inside the stilling well to rise and fall. The stilling well water level is recorded on a chart to provide a time-history of the local sea level fluctuations. The small orifice in the stilling well is designed to filter out wave motion having periods typically smaller than a minute or so. This is accomplished through frictional dissipation at the orifice and by the large ratio of stilling well cross-sectional area to orifice area. Thus higher frequency water level fluctuations such as wind-generated waves would not be recorded but the longer period water level fluctuations of concern in this chapter would. Records from these gages for many locations along the coast and in bays and estuaries are of great value in the study of longer period waves and other water level fluctuations.

4.1. ASTRONOMICAL TIDE

Astronomical tide theory is presented in great detail in several oceanography and other specialized texts (e.g. Defant, 1961; Neumann and Pierson, 1966; and Macmillan, 1966). This chapter concentrates on those aspects of the tide that are of particular interest to coastal engineers.

The gravitational attraction of the moon and sun on the earth and the equal and opposite centrifugal forces are the primary generating forces of the tide. Although the sun's mass is approximately 2.7×10^7 times that of the moon, the greater proximity of the moon to the earth results in the moon having about twice the sun's gravitational force on the oceans. The tide is a long period surface wave generated by these gravitational and centrifugal forces. Because of its long period, the tide propagates as a shallow water wave (Eq. 2.18) even over the deepest parts of the ocean. As the tide wave propagates onto the continental shelf and into bays and estuaries it is significantly affected by nearshore hydrography, friction, Coriolis acceleration, and resonance effects.

The celerity and thus the time of arrival of the tide at a given location are dependent upon the water depths, which also control the refraction of the tide wave. Converging and diverging shorelines cause the amplitude of the tide to increase and decrease, respectively, due to the increase and decrease of energy per unit crest width. Decreasing water depths as the tide wave shoals will increase the tidal amplitude. Bottom friction, which dissipates wave energy, will cause the amplitude to decrease. Thus in shallow nearshore regions, the tide will travel slower than it does at sea, it will usually have a greater amplitude, and it will behave in a rather complex fashion, particularly in irregular bays and estuaries.

The tide is a wave having very low steepness and thus relatively high reflectivity. At places such as the Bay of Fundy, Nova Scotia, reflection causes resonance (see Section 4.6) and amplification of the tidal amplitude in the Bay. Reflection of the tide wave also increases the complexity of the tide in some coastal regions.

In the northern hemisphere, Coriolis acceleration will deflect flowing water to the right (left in the southern hemisphere). Thus as the tide propagates up an estuary causing flow into the estuary, Coriolis acceleration causes the water level to be higher on the right side. On ebb tide, when the flow reverses, the tidal elevation on the opposite side is higher (exclusive of other effects).

Some understanding of the way in which the principal solar and lunar tide generating forces operate may be gained by considering the idealized earth-moon system shown in Fig. 4.2. The two bodies revolve around a

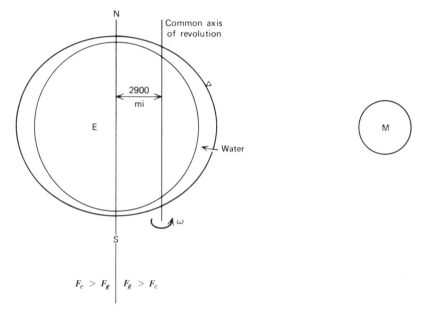

Figure 4.2   Tide generation—idealized earth and moon system.

common axis located about 2900 miles from the center of the earth with the
gravitational F_g and centrifugal F_c forces in balance at the center of mass of
each body. An element of water located on the side of the earth farthest from
the moon would have $F_c > F_g$ and a resulting outward force. On the half of
the earth closest to the moon $F_g > F_c$, which also causes an outward force.
The result is two bulges or two high and two low tides each day as the earth
rotates.

While the earth rotates once, the moon makes $1/29.5$ of a revolution
(lunar month is 29.5 days) so the principal lunar period $M_2 = 12 \, (1 + 1/29.5)$
$= 12.42$ hr. The principal solar period S_2 is 12.0 hr. As the lunar force
predominates, high and low tides progress 0.84 hr (50 min) each day.

The moon and sun are rarely over the equator or in the same plane with
the earth, and the earth's land masses interfere with the progression of the
tidal wave, causing the combined tides due to these principal forces to
become complex. Figure 4.3 shows the approximate orientation of the sun,
moon, and earth at the quarter points of the moon's revolution about the
earth and with reference to the sun. At the first position (new moon) and
third position (full moon) the solar and lunar forces reinforce and the
highest or spring tides occur. At the second and fourth quarters the lowest or

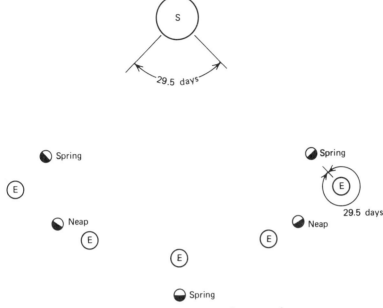

Figure 4.3   Earth-moon-sun, lunar month.

neap tides occur. Each quarter is $29.5/4 = 7.4$ days long, so spring and neap tides are 14.8 days or about two weeks apart. The result of these two principal forces is a water level behavior similar to the beating effect of Fig. 2.7, where the lunar component has a slightly longer period and a greater amplitude than the solar component.

The principal lunar and solar components are just two of over 390 active components having periods ranging from about 8 hr to 18.6 years. Each component has a period that has been determined from astronomical analysis and a phase angle and amplitude that depend on local conditions, and are best determined empirically. Eight of the major components with their common symbol, period, relative strength, and description are listed in Table 4.1. The components combine in different ways at each location and are affected by local hydrography, friction, resonance, and so on, to produce the local tide.

The common types of tides are demonstrated by Fig. 4.4. They range from the semidiurnal type, where the semidiurnal components $M_2 + S_2$ are stronger than the diurnal components $K_1 + O_1$, to the diurnal type where $K_1 + O_1 > M_2 + S_2$. Also, note the spring and neap cycles.

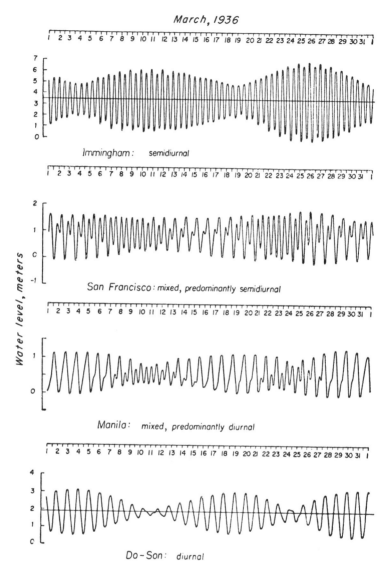

Figure 4.4 Examples of common types of tide (German Tide Tables, 1940).

TABLE 4.1. EIGHT MAJOR TIDAL COMPONENTS

	Symbol	Period (hr)	Relative Strength	Description
Semidiurnal tides	M_2	12.42	100.0	Main lunar seimdiurnal component
	S_2	12.00	46.6	Main solar semidiurnal component
	N_2	12.66	19.1	Lunar component due to monthly variation in moon's distance from earth
	K_2	11.97	12.7	Soli-lunar constituent due to changes in declination of sun and moon throughout their orbital cycle
Diurnal tides	K_1	23.93	58.4	Soli-lunar component
	O_1	25.82	41.5	Main lunar diurnal component
	P_1	24.07	19.3	Main solar diurnal component
Long-period tides	M_f	327.86	17.2	Moon's biweekly component

Coastal water elevations are referenced to a variety of tidal datums in different regions of the world. Some of these datums and other reference water levels are:

1. *Mean sea level* (MSL). The average level of the sea surface.
2. *Mean low water* (MLW). The average of all the low water levels (i.e. every low tide level). MHW is the average of the high tide levels.
3. *Mean lower low water* (MLLW). The average of only the alternate lower of low water levels (see mixed tides in Fig. 4.4). MHHW is the average of the alternate higher high tide levels.
4. Mean tide level (MTL). Level located midway between MLW and MHW.

MSL, MLW and MLLW are usually determined from 19-year records. Published tidal elevation data and hydrographic charts for the Atlantic and Gulf coasts are usually referenced to MLW; on the Pacific coast they are referenced to MLLW. Land elevations are usually referenced to MSL so care must be taken in combining topographic and hydrographic data. One reason for using MLW or MLLW datums for hydrographic charts is to minimize the possibility of navigators running aground at low tide.

Tide ranges and types at many coastal locations in the United States are listed in Table 4.2. The ranges are the difference between MHW and MLW for the Atlantic and Gulf regions and MHHW and MLLW for the Pacific. There is a rather steady progression of tide ranges along the coast but tide ranges in bays and estuaries can show significant local variations.

Tidal fluctuations can generate offshore currents having velocities that are typically less than 2 knots with rotary direction patterns. That is, the current flows continuously with the direction changing cyclically through all points of the compass during a tidal period. When tidal motion is constricted at harbor and estuary entrances, coastal inlets, and so on, reversing flows with rather large velocities can be generated. As examples, at the entrance to San Francisco Bay and in the Cape Cod Ship Canal up to 6-knot tidal currents occur. Tidal current tables for many United States coastal locations are published annually by the National Ocean Survey of the U.S. Department of Commerce.

TABLE 4.2   SELECTED UNITED STATES COASTAL TIDE RANGES

Location	Tide Range (m)	Type of Tide
Liverpool Bay, Nova Scotia	1.4	Semi-diurnal
Boothbay Harbor, Me.	2.9	Semi-diurnal
Boston, Mass.	3.1	Semi-diurnal
Fire Island, LI, N.Y.	1.4	Semi-diurnal
Sandy Hook, N.J.	1.5	Semi-diurnal
Ocean City, Md.	1.1	Semi-diurnal
Virginia Beach, Va.	1.1	Semi-diurnal
Cape Fear, N.C.	1.5	Semi-diurnal
Myrtle Beach, S.C.	1.7	Semi-diurnal
Savannah Riv. Ent., Ga.	2.3	Semi-diurnal
Palm Beach, Fla.	0.9	Semi-diurnal
Key West, Fla.	0.4	Semi-diurnal
Pensacola, Fla.	0.4	Diurnal
Mobile, Ala.	0.5	Diurnal
Galveston, Tex.	0.5	Diurnal
La Jolla, Cal.	1.2	Mixed
San Francisco, Cal.	1.3	Mixed
Crescent City, Cal.	1.7	Mixed
Columbia River Entrance	1.8	Mixed
Grays Harbor, Wash.	2.3	Mixed
Juneau, Ak.	4.5	Mixed
Dutch Harbor, Ak.	0.7	Mixed
Honolulu, Ha.	0.4	Mixed

4.2. TIDE PREDICTION

A long record of the tide measured on a continuous, or at least hourly, basis by a tide gage such as shown in Fig. 4.1, or by some other device such as a submerged pressure sensor designed to dampen high frequency fluctuations is needed to analyze and then predict the tide at a particular location. Ideally, the record should be 19 years long but records as short as 370 days can be used.

The instantaneous tidal elevation η above a selected datum can be given by

$$\eta = A + \sum_{i=1}^{N} A_i \cos\left(\frac{2\pi t}{T_i} + \Delta_i\right) \dots \qquad (4.1)$$

where: $A =$ vertical distance between the datum and MSL; A_i, T_i, and Δ_i are the amplitude, period, and phase angle of a particular component (e.g. M_2, K_1), t is time, and N is the number of components used. The tide record is analyzed (Schureman, 1940) to evaluate A_i and Δ_i for each component of known period T_i. Although the eight components in Table 4.1 are sufficient to analyze the tide at many locations, the National Ocean Survey considers the 37 most important components in performing this analysis. Once A_i and Δ_i are known for each component, Eq. 4.1 can be used to predict future tide levels at the location where the analyzed record was measured. This can be done mechanically but is now usually done by computer. The results are published annually by the National Ocean Survey for the U.S. coasts and many other locations around the world. Data are presented in terms of the elevation and time of high and low tide levels, and a technique for constructing the complete tide curve is presented. Remember, this is the predicted astronomical tide and does not include meterological effects that may be active at a particular time.

4.3. TSUNAMIS

The term "tsunami" is used to denote relatively long period waves generated by coastal and undersea seismic disturbances (earthquakes) and related landslides, bottom slumping, and volcanic eruptions. Although tsunami waves have a low amplitude at sea, shoaling, refraction and resonance can greatly increase the nearshore amplitude and onshore runup of these waves. This has caused several major catastrophies including the loss of many lives in coastal areas prone to tsunami attack. A thorough discussion of tsunamis is given by Wiegel (1970) and an informative and detailed analysis of the 1964 Alaskan earthquake and resulting tsunami is given by Wilson and Tørum (1968).

Tsunamis are generated by a rapid large-scale disturbance of a mass of ocean water that results in a displacement of the ocean surface and the creation of waves. Tsunami generation usually requires sea bed movement having a significant vertical component in sufficiently shallow water. Most recorded tsunamis have been generated by earthquakes having a focal depth of less than 60 km and a magnitude of 6.5 or higher on the Richter scale (Iida, 1969). The 1964 Alaskan tsunami was generated by an earthquake of 8.4–8.6 magnitude (aftershocks up to 6.5) and 20–50 km focal depth that resulted in land movement over an 800 km front in a period of 4–5 min (Wilson and Tørum, 1968). The sea bed rotated around a hinge line with underwater uplift and subsidence in excess of $+8$ m and -2 m, respectively. Most tsunamis are generated at the active earthquake regions along the rim of the Pacific Ocean (Aleutian Islands, Japan, New Zealand, and the west coast of South America) although weaker tsunamis have been recorded in other parts of the world.

A tsunami will typically consist of a group of waves having periods of 5–60 min and irregular amplitudes. Assuming a typical period of 20 min and using the mean ocean depth of 3800 m, from Eq. 2.18

$$C = \sqrt{gd} = \sqrt{9.81\ (3800)} = 193 \text{ m/sec } (432 \text{ mph})$$

$$L = CT = 193\ (20)\ 60 = 2.31 \times 10^5 \text{ m } (144 \text{ miles})$$

Since $d/L = 0.016 < 0.05$ the use of Eq. 2.18 is justified. Even a 5-min period wave traveling over the deeper ocean depths (about 6000 m) behaves essentially as a shallow water eave. Most authors (see Wiegel, 1970) agree that in the open ocean tsunami wave heights are of the order of a meter or less.

Rather accurate tsunami wave refraction diagrams can be constructed using the techniques presented in Section 3.2 (with $C = \sqrt{gd}$) if sufficient information about the generation region and initial wave crest orientation is available. Map projection distortion must be considered and bottom irregularities much shorter than a wave length should be smoothed. Keulegan and Harrison (1970) present a technique for constructing tsunami refraction diagrams by digital computer. Refraction diagrams were constructed for tsunamis generated in Alaska; Kamchatka, Russia; and Chile and propagating to Hilo, Hawaii and Crescent City, California. A major objective was to determine nearshore wave crest orientations for use as input to hydraulic model studies of proposed tsunami barrier plans.

The travel time t_T of a tsunami wave from its source point to a point of interest can be predicted by summing the travel times along successive increments of the wave orthogonal connecting the two points. Thus

$$t_T = \sum \frac{\Delta s}{\sqrt{gd_s}}$$

where d_s is the average water depth over the increment of length Δs. Green (1946) calculated travel times for the 1946 Aleutian tsunami by summing along great circle routes rather than along orthogonals from a refraction diagram. For a point 1100 miles and 3 hr from the source, the travel time was predicted within a few minutes, while for a point 8000 miles and 18 hr away, the travel time was predicted within a half hour. For places (such as Hawaii) frequently attacked by tsunamis, charts of travel time from all possible source areas have been developed (see Zetler, 1947 and Gilmour, 1961). These charts are applicable for any tsunami wave since tsunami celerity only depends on water depth.

4.4. NEARSHORE CHARACTERISTICS OF TSUNAMIS

An indication of the changes that occur as tsunami waves approach shore can be obtained by continuing the example calculation for a 20-min wave in, say, 10 m water depth. Thus

$$C = \sqrt{9.81\ (10)} = 9.9 \text{ m/sec } (22.1 \text{ mph})$$

$$L = 9.9\ (20)\ 60 = 1.19 \times 10^3 \text{ m } (7.4 \text{ miles})$$

From Eq. 2.40 (ignoring refraction and friction), if the wave height in 3800 m of water is 1 m, the height in 10 m of water, H_{10}, is

$$H_{10} = \sqrt{\frac{144}{7.4}} = 4.4 \text{ m}$$

since $n = 1.0$ throughout. For a tsunami wave in very shallow water, d in Eq. 2.18 more correctly (see Section 2.2) represents the water depth plus the wave crest amplitude (which is approximately the wave height). Thus a 1 m high wave would become approximately 4.4 m high in 5.6 m of water depth.

Refraction can cause nearshore wave heights to be significantly greater than indicated by the sample calculation. Diffraction causes a spreading of wave energy as tsunami waves propagate across the ocean, so regions closer to the tsunami source are likely to have higher nearshore waves. Large islands such as Hawaii usually have a marked reduction in wave height on the leeward side due to the effects of refraction and diffraction. As tsunami waves are shallow water waves, $C_g = C$ and wave dispersion due to group effects is not important. However, tsunami wave groups will generally increase in complexity with distance from the source due to wave reflection, development of multiple wave trains due to refraction and diffraction, and shelf and embayment resonant oscillations. Water level records for different tsunamis at the same coastal location show greater similarity than records for the same tsunami at different locations. This indicates the importance of local effects on tsunami characteristics.

Figure 4.5, a tide gage record from San Francisco, California, taken after the 1964 Alaskan earthquake, shows the resulting tsunami waves superimposed on the tide. It appears that the tsunami set San Francisco Bay into resonant oscillation to cause noticeable water level oscillations of about 35-min period, which continued for over 24 hr. The earthquake occurred at 0344 hr. The predicted arrival time at San Francisco (Spaeth and Berkman, 1965), which was broadcast at 0530 hr, was 0915 hr. The first crest arrived at about 0900 hr.

Tsunami waves have an extremely low steepness and, as a result, a high relative runup. Although Fig. 2.12 does not cover the range of typical tsunami wave steepnesses (H_0'/T^2 in the order of 10^{-7}) it does indicate that, even for flat slopes, the relative runup can be quite high. Kaplan (1955) conducted wave tank experiments on the runup of low steepness waves on $1:30$ and $1:60$ plane slopes in order to investigate tsunami runup. He

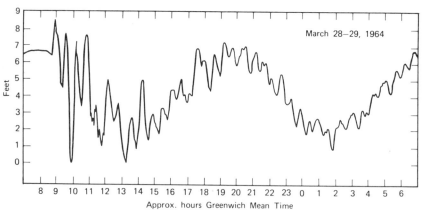

Figure 4.5 Tide gage record at San Francisco (from Spaeth and Berkman, 1965).

presented the empirical equations

$$(1:30 \text{ slope}) \quad \frac{R}{H} = 0.381 \left(\frac{H}{L} \right)^{-0.316} \tag{4.2a}$$

$$(1:60 \text{ slope}) \quad \frac{R}{H} = 0.206 \left(\frac{H}{L} \right)^{-0.315} \tag{4.2b}$$

where H and L are the incident wave height and length at the toe of the slope and R is the vertical runup elevation above the still water level. His data were for H/L values greater than 10^{-3}. Continuing the example, a 4.4 m high wave in 5.6 m water depth has $H/L = 3.7 \times 10^{-3}$ and, from Eq. 4.2a for a $1:30$ slope, $R/H = 3.6$ and $R = 15.9$ m. For the 1946 Aleutian tsunami, Van Dorn (1965) reports runup heights in the order of 15 m at several locations on the island of Hawaii. He also found that the coastwise distribution of wave runup values varied by a factor of 10 due to local effects and that the mean runup on the incident side of the island was three times that on the leeward side.

Thus tsunami runup can be very high and quite irregular at different points along the same coastline for a particular tsunami. The best approach for predicting possible runup at a site for design purposes is to analyze historical runup data. Distribution curves that give the percent chance of runup to a particular height or higher occurring in a selected time interval (i.e. return period) are developed. Cox (1964) and Wiegel (1970) discuss the development of these curves for Hilo, Hawaii, and Crescent City, California.

In addition to predicting tsunami runup for planning and design purposes, it is important to be able to warn those in the path of a tsunami of its existence and estimated time of arrival. All coastal earthquakes do not generate tsunamis. When an earthquake occurs in the Pacific area, observatories report seismic data so the epicenter can be located. When the existence and severity of a tsunami are reported by a station near the epicenter, previously calculated travel times are used to forecast the arrival time at points along the tsunami's path. This information is broadcast at these points along with appropriate warnings and guidance (Roberts, 1950).

Magoon (1965) surveyed the structural damage along the coast of California due to the 1964 Alaska tsunami. He found the damage at Crescent City, which was particularly hard hit, to be $11 million; the wave runup reached 6.4 m above MLLW. Usually, damage is due to flooding, to the high flow velocities in the runup surge, and to the impact of solid objects carried by the surge.

The velocity u of a surge on a dry bed is given by

$$u = k\sqrt{gd} \qquad (4.3)$$

where d is the water depth near the front of the surge and k is a coefficient varying from about 0.7 for high bed resistance to 2.0 for a frictionless bed (Wilson and Tørum, 1968). Assuming a typical surge depth of 3 m and $k = 1.5$, Eq. 4.2a yields as surge velocity of 14.1 m/s which is in general agreement with reported values (see Wilson and Tørum, 1968). Velocities of this magnitude can cause major structural damage and can carry off objects such as cars and boats. Cross (1967) presents results of his study of tsunami surge characteristics and resulting pressures and forces on vertical structures.

Some measures that can be taken to contend with tsunami wave attack are: conduct evacuation drills including the removal of vessels from harbors; remove structures that are weak or easily damaged by flooding; design structures (geometry, orientation, strength) to withstand tsunami surge; plant groves of trees to reduce surge velocities and total runup; and construct offshore barriers such as the one proposed for Hilo, Hawaii (see Kamel, 1970).

4.5. BASIN OSCILLATIONS

There are systems that will respond to a disturbance by developing a restoring force that restores the system to its equilibrium condition. Inertia carries the system past the equilibrium condition and a free oscillation at the system's natural period or frequency is established. A classic example of one such system is a simple pendulum. The natural frequency of oscillation depends on the geometry of the system. It is essentially independent of the magnitude of the initial disturbance, which does, however, establish the oscillation magnitude of the system. After the initial disturbance has occurred, free oscillations continue at the natural frequency but with exponentially decaying amplitude due to the effects of friction. These systems can also undergo forced oscillations at frequencies other than the natural frequency owing to a cyclic energy input at non-natural frequencies. Continuous excitation at frequencies equal or close to the natural frequency will usually cause an amplified system response, the level of amplification depending on the proximity of the excitation frequency to the natural frequency and on the frictional characteristics of the system.

An enclosed or partly enclosed basin of water such as a lake, bay, or harbor can be set into free oscillation at its natural frequency (and harmonic modes) or forced oscillation, as described above. The result is a surging or seiching action of the water mass as wave motion propagates back and forth across the basin. The speed and direction of wave propagation and the resulting natural frequency of the basin depend on the basin geometry. Typical sources of excitation energy include:

1. Ambient wave motion if the basin has an opening (e.g. harbor open to the sea) to permit entry of this energy.

2. Atmospheric pressure fluctuations.

3. Tilting of the water surface by wind stress, with subsequent release.

4. Local seismic activity.

5. Eddies generated by currents moving past the entrance to a harbor.

Oscillations of bays and harbors are usually of low amplitude and relatively long period. Their damage potential is due primarily to (1) the large scale horizontal motions which can adversely affect moored vessels (vessels collide, mooring lines break, fender systems are damaged, loading operations are delayed, etc.) and (2) the strong reversible currents that can be generated at harbor entrances and other constricted flow regions and that can be detrimental to navigation.

Additional insight to the nature of resonance, resonant amplification, and the phase relationship between the excitation and system response can be had by considering the behavior of a single degree of freedom, linearly damped, vibrating spring-mass system (see Raichlen, 1966 and Wilson, 1972 for analytical development). For example, a mass hanging on the end of a spring and suffering various degrees of damping as it oscillates will behave as indicated by the curves of Fig. 4.6. A is the amplification or ratio of mass displacement to excitation displacement, T is the excitation (and response) period, T_n is the natural period of the mass-spring system, and ϕ is the phase angle by which the mass displacement lags the excitation displacement.

Excitation at periods much greater than the natural period ($T_n/T \cong 0$) results in a system response having approximately the same magnitude and phasing as the excitation force. In simpler words, the mass responds directly to the excitation. As the excitation period decreases toward the natural period of the system, the mass response is amplified and a phase lag develops. The amount of amplification and phase lag depend on the friction and the ratio T_n/T. When $T_n = T$ the amplification is greatest and the phase lag is 90°. At this point the mass velocity, which is 90° out of phase with the displacement, is in phase with the excitation force. Note that frictional effects cause a slight decrease in the natural period of the system. At $T < T_n$ the amplification continually diminishes and the phase lag approaches 180°.

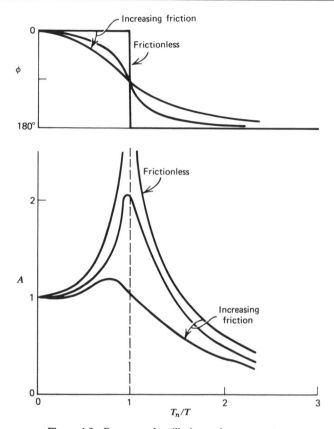

Figure 4.6 Response of oscillating spring-mass system.

If the mass is extended and released rather than being cyclically excited, it would oscillate at its natural period T_n. With a cyclic excitation at the natural period of the system, the response amplitude grows to the equilibrium value given by Fig. 4.6. At this point the rate of energy input equals the rate of energy dissipation by friction. Thus if the system was frictionless, amplification would be infinite. When the excitation is removed, friction causes the response amplitude to decrease exponentially with time.

The concepts demonstrated by Fig. 4.6 apply to a basin of water, except it will typically have several degrees of freedom, that is, several modes of oscillation and resulting natural frequencies and harmonics. In some cases, such as the excitation of a bay by tsunami waves, the duration of excitation may not be sufficient to develop a fully amplified bay response. If a harbor is excited by a spectrum of wave energy, it will selectively amplify those periods around the natural period as demonstrated by Fig. 4.6. There will usually be a free oscillation at the natural period and forced oscillations at surrounding periods.

4.6. RESONANT MOTION IN TWO-DIMENSIONAL BASINS

In this section evaluation of the water motion for the fundamental and harmonic modes of free oscillation and the resonant periods for these modes will be considered. Figure 4.7 shows the fundamental and first and second harmonic modes of oscillation in idealized closed and open two-dimensional basins. Details of the water motion characteristics have been discussed in Section 2.7 on wave reflection. The water-surface time-history forms an antinode at the vertical barriers and a node at the opening to large (ideally infinite) seas. In each case the basin length and the wave length have a fixed

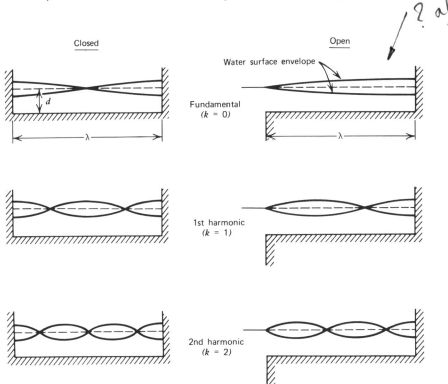

Figure 4.7   Water surface profiles for oscillating basins.

relationship, so the incident and reflected waves reinforce to produce a stable standing wave pattern. Since the period for a particular mode of oscillation is equal to the wave length divided by the wave celerity, and most basin oscillations of engineering concern involve shallow water waves $(C = \sqrt{gd}\,)$,

$$T_n = \frac{2\lambda}{(k+1)\sqrt{gd}} \qquad \text{(closed basin)} \qquad (4.4)$$

and

$$T_n = \frac{4\lambda}{(2k+1)\sqrt{gd}} \qquad \text{(open basin)} \qquad (4.5)$$

Here, k is as given in Fig. 4.7 and λ and d are the basin length and depth, respectively. The fundamental mode of oscillation has the longest period and the harmonic periods decrease from this by factors of $(k+1)^{-1}$ and $(2k+1)^{-1}$ for closed and open basins.

The maximum horizontal water excursion and particle velocities occur below nodal points. Equations that indicate their magnitude can be developed from Fig. 4.8, which shows a section of a standing wave. If one assumes the water surface profile to be sinusoidal, the volume of water $\forall$ that must flow across a vertical line through a nodal point in a half period (shaded area) is

$$\forall = 2 \int_{L/4}^{L/2} \frac{H}{2} \sin\left(\frac{2\pi x}{L}\right) dx = \frac{HL}{2\pi} \qquad (4.6)$$

where H is the standing wave height and L the length. The time average horizontal velocity $\bar{V}$ under the nodal point is the volume of water divided

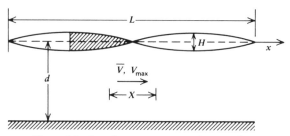

Figure 4.8 Standing wave water surface envelope.

by the time $T_n/2$ and cross-section area d or

$$\bar{V} = \frac{HL}{\pi \, dT_n} \tag{4.7}$$

Since the motion is sinusoidal, the maximum horizontal velocity V_{max} is $\dfrac{\pi \bar{V}}{2}$ or

$$V_{max} = \frac{HL}{2 \, dT_n} = \frac{HC}{2d} = \frac{H}{2}\sqrt{\frac{g}{d}} \tag{4.8}$$

The horizontal particle excursion at the node X is the average velocity times the half period or

$$X = \frac{\bar{V} T_n}{2} = \frac{HL}{2\pi d} = \frac{HT_n}{2\pi}\sqrt{\frac{g}{d}} \tag{4.9}$$

Considering a rectangular section of a closed basin with, say, $d=8$ m and $\lambda = 1000$ m and a typical oscillation height of 0.2 m, Eq. 4.4 with $k=0$ yields

$$T_n = \frac{2\,(1000)}{\sqrt{9.81(8)}} = 226 \text{ sec } (3.77 \text{ min.})$$

and Eqs. 4.8 and 4.9 yield

$$V_{max} = \frac{0.2}{2}\sqrt{\frac{9.81}{8}} = 0.11 \text{ m/sec}$$

and

$$X = \frac{0.2\,(226)}{2\pi}\sqrt{\frac{9.81}{8}} = 7.96 \text{ m}$$

Although the maximum velocity is low, the long period and large horizontal motion could cause serious difficulties to moored vessels as discussed in Section 4.5.

Many natural basins have a much greater length than width and the oscillations that develop in the longitudinal direction are more common and of greater engineering importance than those in the transverse direction. Equations 4.4 and 4.5, in modified form, can be used to predict the periods of free oscillation in the longitudinal direction for narrow basins with irregular cross-sections and centerline profiles. The speed of a shallow water wave at a particular cross-section normal to the direction of wave propagation can be calculated by using the hydraulic depth (Chow, 1959), which equals the cross-sectional area divided by the width at the free surface. Equations 4.4 and 4.5 become

$$T_n = \frac{2}{(k+1)} \int_0^\lambda \frac{dx}{\sqrt{gd}} \qquad \text{(closed basin)} \qquad (4.10)$$

and

$$T_n = \frac{4}{(2k+1)} \int_0^\lambda \frac{dx}{\sqrt{gd}} \qquad \text{(open basin)} \qquad (4.11)$$

where the hydraulic depth at any cross-section d is a function of the distance x from the end of the basin. Usually, the relationship between d and x can not be put into equation form for the solution of Eqs. 4.10 and 4.11. However, a numerical calculation can be made using

$$T_n = \frac{2}{(k+1)} \sum_{i=1}^N \frac{\Delta x_i}{\sqrt{gd_i}} \qquad \text{(closed basin)} \qquad (4.12)$$

$$T_n = \frac{4}{(2k+1)} \sum_{i=1}^N \frac{\Delta x_i}{\sqrt{gd_i}} \qquad \text{(open basin)} \qquad (4.13)$$

where the irregular bottom is replaced by N uniform depth segments of length Δx_i and hydraulic depth d_i.

More sophisticated numerical techniques that give the relative water surface displacement along the basin axis as well as a more accurate determination of the periods of oscillation are available. For two of these the reader is referred to Defant (1961) and Raichlen (1966). The periods of oscillation for several simple geometric profile and plan configurations, derived from the basic equations of continuity and motion, have been summarized by Wilson (1972). Many of these shapes approximate natural basins and thus may be used in an analysis of resonant frequencies.

The preceeding discussion assumed each boundary is either closed, and thus an antinode, or completely open to an infinite sea, and thus a node. However, many boundaries have partial openings and/or are open to a finite body of water. The resulting boundary condition is more complex; it causes resonant behavior at the opening that is somewhat between that of a node and antinode, and the basin resonant frequencies are commensurately modified.

4.7.   RESONANT MOTION IN THREE-DIMENSIONAL BASINS

Basins having widths and lengths of comparable size can develop more complex patterns of resonant oscillation. The character of these oscillations can be demonstrated by the equations for the periods and water surface configurations for resonant oscillations in a rectangular basin (Fig. 4.9). Only an outline of the derivation will be presented; see Raichlen (1966) for more of the details.

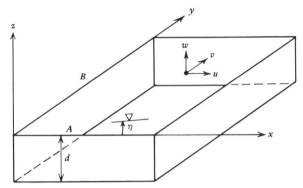

Figure 4.9 Definition sketch for three-dimensional basin oscillations.

A solution for the free oscillations must satisfy the three-dimensional Laplace equation,

$$\frac{\partial^2 \phi}{\partial x^2} + \frac{\partial^2 \phi}{\partial y^2} + \frac{\partial^2 \phi}{\partial z^2} = 0$$

and the following boundary conditions at the water surface, side walls, and bottom, respectively (see Section 2.1):

$$\eta = -\frac{1}{g}\frac{\partial \phi}{\partial t} \qquad |z=0$$

$$u = \frac{\partial \phi}{\partial x} = 0 \qquad |x=0,A$$

$$v = \frac{\partial \phi}{\partial y} = 0 \qquad |y=0,B$$

$$w = \frac{\partial \phi}{\partial z} = 0 \qquad |z=-d$$

Assuming shallow water conditions the solution is

$$\phi = \frac{Hg}{2\sigma}\cos\frac{n\pi x}{A}\cos\frac{m\pi y}{B}\sin\sigma t \qquad (4.14)$$

where n, m (integers $0,1,2,3\dots$) define the various modes of oscillation. The water surface time-history is

$$\eta = \frac{H}{2}\cos\frac{n\pi x}{A}\cos\frac{m\pi y}{B}\cos\sigma t \qquad (4.15)$$

Application of the continuity equation leads to the equation for the fundamental and harmonic periods of oscillation

$$T_{n,m} = \frac{2}{\sqrt{gd}}\left[\left(\frac{n}{A}\right)^2 + \left(\frac{m}{B}\right)^2\right]^{-1/2} \qquad (4.16)$$

Note, for a long narrow basin with longitudinal oscillations ($m=0, n=1,2,3\dots$) Eq. 4.16 reduces to

$$T_n = \frac{2A}{n\sqrt{gd}}$$

which is the same as Eq. 4.4.

At nodal points, $\eta = 0$ for all values of t and from Eq. 4.15,

$$\cos\frac{n\pi x}{A}, \cos\frac{m\pi y}{B} = 0$$

thus

$$\frac{n\pi x}{A}, \frac{m\pi y}{B} = \frac{\pi}{2}, \pi, \dots$$

and

$$x = \frac{A}{2n}, \frac{A*}{n}, \dots$$

$$y = \frac{B}{2m}, \frac{B*}{m}, \dots$$

*The number of terms to be used is the number of nodes given by the values of n and m.

As an example, consider a square basin $(A = B = \lambda)$ with $n = m = 1$. Then

$$\eta = \frac{H}{2} \cos\frac{\pi x}{A} \cos\frac{\pi y}{B} \cos\frac{2\pi t}{T}$$

and

$$T = \frac{\sqrt{2}\,\lambda}{\sqrt{gd}}$$

The water surface pattern at $t = 0$ is as shown in Fig. 4.10 where the water surface contours above the still water level are solid and those below are dashed. At $t = T/4$ the surface is flat; at $t = T/2$ the solid and dashed contours are reversed; at $t = 3T/4$ the surface is flat again; and at $t = T$ the surface is back to the way it was at $t = 0$. The maximum water level range H is found at the four corners, which are antinodes. For a particular amplitude of oscillation the water particle velocities may be evaluated from Eq. 4.14

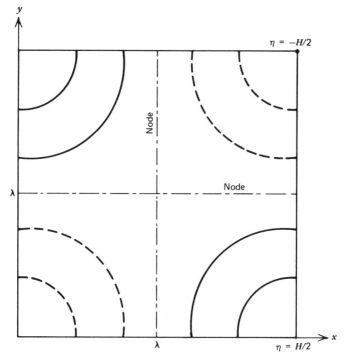

Figure 4.10 Water surface contours, rectangular basin, $n,m = 1$, $t = 0, T$.

and $u = \partial\phi/\partial x, v = \partial\phi/\partial y$ and $w = \partial\phi/\partial z$. The reader is encouraged to work out the patterns for other modes of oscillation (e.g. $n = 2$, $m = 1$; $n = 3$, $m = 2$).

Wilson (1962) reports a study of resonant oscillations at Duncan Basin, Cape Town, South Africa. This is a rectangular harbor 1825 m long, 640 to 670 m wide, and 13 m deep with a 122 m navigation opening on one of the long sides. The harbor behaves as a closed basin oscillating primarily at the $n = 0$, $m = 1$; $n = 1$, $m = 0$; and $n = 2$, $m = 0$ modes, which have 5.6-, 1.8-, and 0.9-min periods, respectively. Apparently the harbor responds to subharmonics of longer period coastal oscillations generated by moving storm systems. The opening was initially 228 m wide but harbor oscillations were so undesirable that the opening width was decreased. Due to a subsequent increase in ship traffic it was proposed to widen the entrance from 122 to 183 m. A hydraulic model study showed that the amplitudes of oscillation would increase 30 to 35 percent in important sections of the harbor. Model studies were also conducted to investigate possible reductions in amplification achieved by relocating the harbor entrance and by installing a pier normal to one of the shorter ends of the harbor. In each case, the patterns of harbor response were significantly changed but the peak oscillations continued to exceed permissable levels.

The periods and nodal line positions for the various modes of oscillation of a harbor depend only on the geometry of the harbor. This information, and to a large degree the relative amplitudes of horizontal and vertical water displacement and velocity, can be investigated by analytical and physical model techniques. However, the absolute magnitudes of water displacement and velocity depend on the amplitude and duration of the excitation wave energy and the dissipation characteristics of the harbor, which can only be predicted by a field monitoring program.

Many basins have planform and bottom geometries too irregular to be approximated by the results of analytical solutions. One approach to handling basins of this type is the hydraulic model designed on the basis of Froude number similarity. If the model to prototype horizontal and vertical scale ratios are the same (undistorted model) this leads to

$$\left(\frac{\lambda/T}{\sqrt{g\lambda}} \right)_r = 1$$

or

$$T_r = \sqrt{\lambda_r} \tag{4.17}$$

where T is time (e.g. period of oscillation), λ is the length, and the subscript r

denotes the model to prototype ratio. Thus if a model basin is constructed at some constant model to prototype scale (λ_r), the measured resonant periods can be converted to prototype values by Eq. 4.17. The oscillation patterns (nodal and antinodal lines) and the relative amplification of oscillations at the various modes will also be given (as discussed in the previous paragraph).

If the horizontal and vertical scale ratios are not equal (distorted model), as is often necessary, Froude number similarity leads to

$$T_r = \left(\frac{\lambda_h}{\sqrt{\lambda_v}} \right)_r \tag{4.18}$$

where the subscripts h and v refer to the horizontal and vertical scale ratios, respectively. Biesel (1955) and Raichlen (1966) discuss some of the problems common to the conduct of hydraulic model studies of basin oscillation.

Several finite difference numerical schemes have also been developed to investigate resonant oscillations in irregular basins. Some of these are summarized by Wilson (1972). They essentially involve dividing the basin with a horizontal grid system into square or rectangular elements and writing the equations of continuity and motion in finite difference form for each element. The equation matrix can be solved to yield the eigenfrequencies of oscillation and relative surface elevations. Or, if the periods of oscillation are known or determined by trial and error, the equations for each element may be solved in an element-by-element fashion in time and space to evaluate the surface elevations and velocities throughout the basin as a function of time. In setting up the equations for each element, antinodal boundaries have no flow and nodal boundaries have a constant surface elevation throughout time. As discussed at the end of Section 4.6, at openings to other noninfinite basins the boundary is neither a node nor antinode, so the surface amplitude and slope for each side of the boundary must be matched.

4.8. HELMHOLTZ RESONANCE

In addition to the standing wave modes of oscillation discussed in Sections 4.5 and 4.6, a basin open to the sea through an inlet can resonate in a mode known as the Helmholtz mode. Motion is analogous to that of a Helmholtz resonator in acoustics. The water surface in the basin uniformly rises and falls while the inlet channel water mass oscillates in and out. Motion is also analagous to a single degree of freedom spring-mass system with the inlet channel water representing the mass and the basin water level under the action of gravity representing the spring. The resonant period T_H is given (Carrier et al., 1971) by

$$T_H = \sqrt{\frac{(L_c + L_c')A_b}{gA_c}} \tag{4.19}$$

where A_b is the basin surface area, A_c is the channel cross-section area, L_c is the channel length and L_c' is an additional length to account for the mass outside each end of the channel that is involved in the resonant oscillation. L_c' is given by (adopted from Miles, 1948)

$$L_c' = \frac{-W}{\pi} \ln\left(\frac{\pi W}{\sqrt{gd_c}\, T_H}\right) \tag{4.20}$$

where W and d_c are the channel width and depth, respectively.

The Helmholtz mode, which has a period that is greater than the fundamental mode given by Eq. 4.4 or 4.5 seems to be important for ocean harbors responding to tsunami excitation (Miles, 1974). It is also the most significant mode of oscillation for several harbors on the Great Lakes (Sorensen and Seelig, 1976) that respond to the storm-generated long wave energy spectrum on the Lakes.

4.9. STORM SURGE

A storm over nearshore waters can generate large water level fluctuations if the storm is sufficiently strong and the nearshore region is shallow over a large enough area. This is commonly known as storm surge or the meteorological tide. Storm activity can cause both a rise (set up) and fall (set down) of the water level at different locations and times, with the set up predominating in magnitude, duration and areal extent. Specific causes of water level change include: surface wind stress (and related bottom stress due to currents generated by the surface wind stress); Coriolis acceleration; atmospheric pressure differentials; wind wave set up; long wave generation by the moving pressure disturbance; and precipitation and surface runoff.

Storm surge is generally not an important factor in water level analysis on the Pacific coast of the United States. However, on the Gulf and Atlantic coasts where the continental shelf is generally much wider than it is on the Pacific coast and where hurricanes and extratropical storms are common, storm surge is extremely important. Storm-generated water level fluctuations in the shallower of the Great Lakes (especially at the Buffalo and Toledo ends of Lake Erie) can also be significant.

Hurricane Camille in August 1969 had estimated sustained peak wind speeds of 165 knots as it crossed the Mississippi coast. The storm surge reached a maximum of 6.9 m above MSL at Pass Christian, up to 25 cm of rainfall were measured, and the coastal area in the region of highest winds suffered virtually complete destruction (U.S. Army Engineer District, New Orleans, 1970). Surge levels in excess of 3 m above MSL occurred along the coast from the Mississippi River delta to the Mississippi-Alabama line, a coastline distance of over 100 miles. Estimated storm damage (no major urban areas were hit) was just under $1 billion.

Storm surge calculations require a knowledge of the spatial and temporal distribution of wind speed, wind direction, and surface air pressure for the design storm conditions. The Atlantic coast of the United States experiences extratropical storms that result from the interaction of warm and cold air masses and that can generate a substantial storm surge. However, throughout most of the Atlantic coast south of Cape Cod and the Gulf coast, the worst storm condition or design storm will usually be a hurricane of tropical origin.

Design wind and pressure fields for a site may be established by using the measured conditions from the worst storm on record in the general area. Or, if sufficient historical data exist, a return period analysis may be conducted to select storm design parameters having a specified frequency of occurrence for that area. Also, if sufficient historical surge elevation data are available, direct surge-frequency relationships can be developed for a given area. Bodine (1969) did this for the Gulf coast of Texas using data from 19 hurricanes dating from 1900 to 1963.

The U.S. Weather Bureau and U.S. Army Corps of Engineers have developed hypothetical design hurricanes for the Atlantic and Gulf coasts that are based on a return period analysis of significant parameters (Graham and Nunn, 1959). These are known as Standard Project Hurricanes (SPH) and will be discussed in detail in subsequent paragraphs, as most coastal storm surge calculations for design purposes are based on the SPH and the related larger Probable Maximum Hurricane or PMH (U.S. Weather Bureau, 1968).

A hurricane is a cyclonic storm with wind speeds in excess of 75 mph that originates near the equator. Those affecting the United States typically move north west (often with very irregular paths) into the Gulf of Mexico or up the Atlantic coast eventually veering east and out over the Atlantic to dissipate. The driving mechanism is warm moist air that flows toward the eye (center) of the hurricane and gives off heat as it rises and the moisture condenses. After rising in the eye the air flows outward at higher altitudes. The deficiency of warm moist air when the hurricane is over land or colder water at higher ocean latitudes causes the hurricane to dissipate. At ground level, Coriolis acceleration causes the air flow to the eye to take on a counterclockwise (clockwise in southern hemisphere) spiral motion. Wind velocities increase to a maximum V_x (velocity 10 m above sea level) at a radius R a small distance from the eye, and then diminish rapidly to very low magnitudes at the eye. The air pressure continually drops from the ambient pressure at the outer edge to the lowest pressure at the eye. This pressure is known as the Central Pressure Index (CPI) when given in inches of mercury. Other important hurricane parameters are its forward speed V_F and its direction of travel. The reader is referred to the book by Dunn and Miller (1964) for a detailed discussion of hurricane characteristics.

The SPH is a series of hurricanes designed "to represent the most severe combination of hurricane parameters (e.g. CPI, R, V_x) that is reasonably characteristic of a region excluding extremely rare combinations." They were developed from an analysis of data (1900–1956) for hurricanes occurring in four zones along the Atlantic coast and three zones along the Gulf coast. The principal intensity criterion is the CPI, which has a 100 year average frequency of occurrence in a particular zone of interest. For a specific site location one can select a SPH with a small or large R and a high or moderate V_F and related V_x. SPH index characteristics for Galveston, Texas, for example, are (Graham and Nunn, 1959)*:

> Zone C
> CPI = 27.52 in. of Hg
> R(small) = 7 n.m.
>   (medium) = 14 n.m.
>   (large) = 26 n.m.
> V_F (moderate) = 11 knots   (V_x = 101 mph)
>   (high) = 28 knots   (V_x = 111 mph)

For each zone, R, and V_F a surface isovel pattern is given by Graham and Nunn. That for Zones B and C (west Florida to Texas), large R, and high V_F is given in Fig. 4.11. The concentric solid lines give the wind velocity at 10 m above the water surface and the arrows indicate the wind direction. With $R = 26$ n.m. the hurricane force winds (velocity greater than 75 mph) extend over a radius of about 75 n.m. and winds over 40 mph extend over a radius

---

*A subsequent study of hurricanes through 1971 has somewhat revised these index characteristics. See National Weather Service Memo HUR 7-120, June 1972 "Revised Standard Project Hurricane Criteria for the Atlantic and Gulf Coasts."

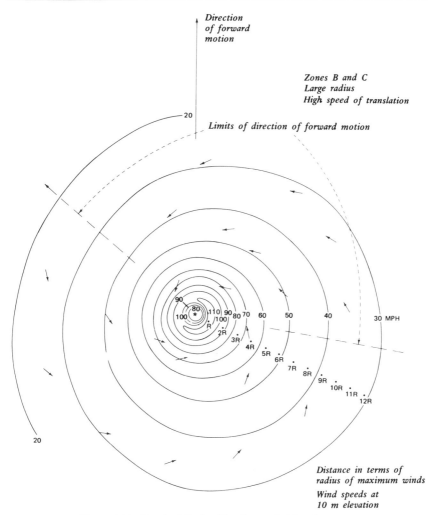

Figure 4.11 Standard Project Hurricane wind isovel pattern.

of about 200 n.m. Note that the wind velocity is higher on the right hand side when looking in the direction of forward motion of the hurricane. Also, owing to the counterclockwise circulation pattern, the onshore wind component is on the right hand side. The result is that the highest surge elevations and worst wind and surge damage typically occur to the right of the eye.

Myers (1954) developed an empirical equation for surface pressure distribution based on an analysis of historical hurricane pressure data. This equation is

$$p_a - p_r = (p_a - \text{CPI})(1 - e^{-R/r}) \qquad (4.21)$$

where p_r is the pressure at any radius r from the eye, and p_a is the ambient pressure, which may be taken as 29.92 in. of Hg for a SPH.

The SPH is adequate for the design of most coastal projects but the advent of nuclear powered electric generating stations in the coastal zone has lead to the development of a Probable Maximum Hurricane. This is "a hypothetical hurricane having that combination of characteristics which will make it the most severe that can probably occur in the particular region involved" (U.S. Weather Bureau, 1968). No frequency of occurrence was assigned to the PMH, which is essentially a thoroughly enlarged SPH. HUR7-97 (U.S. Weather Bureau, 1968) gives detailed procedures for developing the PMH for a particular site.

From economic and other considerations it may be desirable to use a design hurricane other than the SPH or PMH. As examples, a pier with a short design life or a rubble mound breakwater that can easily be repaired should be designed for a storm having a higher frequency of occurrence than the SPH. Such design storms can be developed from the references that present the SPH and PMH, as these give detailed information on the basic hurricane parameters in terms of frequency of occurrence, and they provide procedures for plotting isovel patterns as a function of these parameters. These references also give correction factors to predict the magnitude of diminished wind speeds that occur as a hurricane moves over land.

North of Cape Cod, Massachusetts, the design storm is more likely to be an extratropical cyclone (known as a Northeaster) rather than a hurricane. Peterson and Goodyear (1964) present the characteristics of Standard Project Northeasters that may be used for storm surge prediction in this region.

4.10. STORM SURGE CALCULATIONS

Ideally, storm surge calculations should proceed by solving the unsteady equations of continuity and motion applied to the nearshore water mass with the appropriate surface, bottom, and edge boundary conditions. This can be done in an approximate way with numerical finite difference methods recently developed for digital computers. These models, which involve significant effort, will be briefly discussed at the end of this section. However, as a first step, methods for individually evaluating the setup of each storm surge component will be developed assuming steady state conditions. The setup at the point of interest for each component can then be added to determine the total setup. This approach neglects convective and local water mass accelerations as well as nonlinear interactions among the various components. Also, continuity is not considered. However, this approach can still yield results of sufficient value for many engineering purposes provided it is applied where the assumptions are reasonably satisfied. This approach also gives some insight into the nature of the various storm surge components. The quasi-static bathystrophic approach, which improves somewhat on the separate calculation of individual components, will also be briefly presented.

Initial Setup

It has been observed that the nearshore water level often rises $1/2$ m or more above the astronomical tide level prior to the arrival of a storm. No satisfactory explanation for this phenomena has been presented to date. Marinos and Woodward (1968) found this initial setup to be on the order of 0.6–0.75 meters on the Texas coast, which is probably typical of the entire Gulf coast region. Initial setup is somewhat lower along the Atlantic coast. For calculation purposes some constant initial setup value based on local information from past storms and the magnitude of the design storm should be applied throughout the duration of the storm. One should also select a reasonably large astronomical tide and apply it so that the high tide level coincides with the peak surge level of the storm.

Pressure Setup

From hydrostatics, the water level variation or setup S_p due to a surface pressure variation Δp between two points on a continuous body of water is

$$S_p = \frac{\Delta p}{\rho g}$$

where ρ is the water density. If Δp is the pressure drop from the periphery of a hurricane to a point within the hurricane

$$S_p = \frac{p_a - p_r}{\rho g} = \frac{p_a - CPI}{\rho g}(1 - e^{-R/r}) \tag{4.22}$$

from Eq. 4.21. The pressure setup that occurs at a point of interest as a function of time for a particular hurricane can thus be calculated from Eq. 4.22. For a SPH, the pressure setup at the eye can be as much as one meter of elevation.

Long Wave Setup

A moving surface disturbance will generate waves. These waves achieve their greatest amplitude when the disturbance speed equals the speed of a shallow water wave ($\sqrt{gd}$) and the disturbance has travelled at the required speed for sufficient time for the wave to completely develop (i.e. achieve a balance between rates of energy input and dissipation). Thus a long wave (and resulting water level rise and fall) can be generated by a hurricane moving at the right speed for a sufficient duration. The critical water depths ($V_F = \sqrt{gd}$) for hurricane speeds of 10, 20, and 30 knots are 3, 11, and 26 m, respectively. This phenomenon involves a resonant amplification of the pressure setup previously discussed.

Ewing, Press, and Donn (1954) reported on a squall line ($\Delta p/\rho g \cong 0.03$ m) crossing the lower portion of Lake Michigan and moving at the appropriate speed for over 1/2 hr. A 1-2 m high surge was observed at Michigan City, Indiana, which was in the path of the squall generated wave. A surge of similar peak elevation was observed at the opposite shore about 1 hr later as the wave reflected back across the lake.

There is little information available for predicting the peak amplitude of these long wave surges. From laboratory data, Abraham (1964) suggests the long wave surge elevation (including pressure setup) can approach twice the value given by Eq. 4.22 if a storm moves over water of between 0.75–1.25 times the critical depth for a period of 1 hr or more, and triple that value for a depth range of 0.9–1.1 times the critical depth for 1 hr or more.

Wave Setup

Storm-generated wind waves also cause a nearshore setup S_{ww} due to wave mass transport as discussed in Section 2.6. Equation 2.44 should be used to calculate this component of storm surge from the predicted storm wave climate.

Wind and Bottom Stress Setup

The water surface stress τ_s exerted by the wind is usually given by an equation of the form

$$\tau_s = C_d \rho_a U^2 \qquad (4.23)$$

where ρ_a is the air density, U is the wind velocity at some reference elevation (usually 10 m or 30 ft), and C_d is a drag coefficient that depends upon the surface roughness and related air boundary layer characteristics. Wilson (1960) evaluated and summarized the results of 47 references that presented values for C_d. He concluded that the most reasonable value for C_d varied from 1.5×10^{-3} for light winds asymptotically to 2.4×10^{-3} for strong winds.

It is usually more convenient to write Eq. 4.23 in terms of the water density ρ as

$$\tau_s = k\rho U^2 \tag{4.24}$$

where the wind stress coefficient $k = (1.19 \times 10^{-3})C_d$ (taking $\rho/\rho_a = 845$). A frequently used relationship for k is that given by Van Dorn (1953)

$$k = 1.21 \times 10^{-6} + 2.25 \times 10^{-6}\left(1 - \frac{5.6}{U}\right)^2$$

where U is the wind speed in m/sec. This is based on a field study of a small yacht harbor and gives results slightly higher than Wilson's recommended values.

The surface wind stress generates a current that, in turn, develops a bottom stress, τ_b. Saville (1952), from a field study at Lake Okeechobee, Florida, found that $\tau_s + \tau_b = (3.3 \times 10^{-6})\rho U^2$ and, assuming $k = 3.0 \times 10^{-6}$, suggested that $\tau_b/\tau_s = 0.1$. This ratio has been used for setup calculations if no better data are available. If adequate wind, pressure and resulting setup data are available from a previous storm at a particular site, it is better to determine a local combined surface and bottom stress coefficient K ($\tau_s + \tau_b = K\rho U^2$), which then acts as an "all-inclusive" calibration factor for subsequent calculations.

Figure 4.12 shows a section of the nearshore water column of length Δx, normal to shore, unit width, and depths d and $d + \Delta S_w$. ΔS_w is the setup due to wind and bottom stresses acting over the length Δx. The hydrostatic forces on each end are shown and the bottom stress is drawn in the same direction as the surface wind stress because it is assumed the wind-generated surface current causes a reverse flow on the bottom. Since the water surface slope $\Delta S_w/\Delta x$ is extremely flat, a static balance yields

$$\tau_s \Delta x + \tau_b \Delta x + \tfrac{1}{2} g\rho d^2 - \tfrac{1}{2} g\rho(d + \Delta S_w)^2 = 0$$

Expanding, inserting $\tau_s + \tau_b = K\rho U^2$, and solving

$$\Delta S_w = d\left[\sqrt{\frac{2KU^2\Delta x}{gd^2} + 1} - 1\right] \tag{4.25}$$

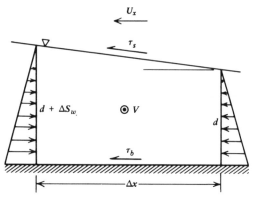

Figure 4.12 Definition sketch for wind and bottom stress and Coriolis setup derivations.

If the wind blows at an angle θ to the x-direction, the effective stress is $(\tau_s + \tau_b)\cos\theta = K\rho U^2 \cos\theta = K\rho U U_x$, where U_x is the component of the wind velocity in the x-direction. Thus Eq. 4.25 becomes

$$\Delta S_w = d\left[\sqrt{\frac{2KU\,U_x\Delta x}{gd^2} + 1} \;-1\right] \tag{4.26}$$

Note that the setup depends on the water depth as does the height and arrival time of the astronomical tide. Thus there is a second order nonlinear coupling of the two that is ignored by evaluating each component separately and adding to obtain the resulting water surface elevation. With a storm surge the astronomical tide will arrive ahead of the predicted time and have a lower amplitude. Bretschneider (1967) suggests the coupled surge level can be as much as 0.3 m below the sum of the individually determined levels.

As a hurricane moves toward the shore, calculations for the setup due to wind and bottom stress can be made by using Eq. 4.26 and assuming that the hurricane remains at a series of positions along the path for a sufficient period of time for steady state conditions to develop at each position. As previously discussed, this approach neglects acceleration effects and is thus of decreasing accuracy as the storm size and forward velocity increase. Also, since continuity of mass is not considered, the hurricane should approach shore in a near normal direction and the shoreline and offshore bottom contours should be reasonably straight.

To initiate the setup calculation, the offshore profile over which the storm traverses is divided into a number of segments (not necessarily equal in size) of length Δx and constant depth d' below MSL. The storm wind field is stationed over the profile at the desired position and the average value of

UU_x is determined for each segment. Proceeding from deep water or the edge of the storm, ΔS_w is calculated for each segment. In this calculation, the depth d in Eq. 4.26 is equal to $d' + \Sigma \Delta S_w$ where $\Sigma \Delta S_w$ is the cumulative setup calculated for the preceeding segments.

Thus d is the total water depth at the seaward end of each segment and it defines the wind and bottom stress surge profile. Similar calculations for each of the sequential storm positions yields the series of wind and bottom stress surge profiles that develop as the storm moves forward.

Equation 4.26 can be applied (with caution) to enclosed bays that are not too irregular in shape. The volume of water that sets up on the leeward end of the bay equals that removed from the windward end and the two sections are divided by a nodal line. For a given storm position the setup calculations proceed in the upwind and downwind directions from the nodal line. Bretschneider (1966) presents a method for making a first estimate of the nodal line location. A balance of the setup and setdown volumes establishes the final surface profile and nodal position.

Coriolis Setup

Coriolis acceleration will cause a moving water mass to deflect to the right (in the northern hemisphere), perpendicular to the direction of motion of the mass. If this deflection is restrained, for example, by a shoreline, as a current moves along the shore with the shoreline to the right, a balance of forces will require a setup of water on the right hand side. In Fig. 4.12 the current of velocity V is coming out of the page and Coriolis acceleration causes a setup as shown. Ignoring water surface and bottom stress a static balance between the hydrostatic forces and the Coriolis acceleration or force per unit mass yields

$$\tfrac{1}{2}\rho g d^2 + \left[2\omega V \sin\phi \right] \rho g d \Delta x - \tfrac{1}{2}\rho g (d + \Delta S_c)^2 = 0$$

The term in brackets is the Coriolis force per unit mass, ϕ is the latitude, ω is the angular speed of rotation of the earth $(7.28 \times 10^{-5}$ rad/sec), and ΔS_c is the Coriolis setup. Neglecting higher order terms,

$$\Delta S_c = \frac{2\omega}{g} V \sin\phi \Delta x \qquad (4.27)$$

Away from the shore, wind-generated currents are free to respond to the Coriolis acceleration; it is only nearshore that this response is restricted and setup (or setdown) develops. It is very difficult to predict the magnitude of nearshore wind-generated currents as well as the degree of restriction and resulting setup that occurs. Bretschneider (1967) solved the equation of motion for the alongshore direction, including only the wind and bottom stresses and the resulting local acceleration, to obtain an equation for longshore current velocity. For the resulting steady state condition (i.e. sufficient wind duration)

$$V = Ud^{1/6}\sqrt{\frac{k}{14.6n^2}}\sin\theta \qquad (4.28)$$

where n is Manning's roughness coefficient (typical value is 0.035), θ is the angle between the wind direction and a line perpendicular to the coast and/or bottom contours, and V is the resulting longshore velocity. The derivation of Eq. 4.28 assumes that there is no flow component normal to the shore, limiting its use to the nearshore region.

The obvious difficulties involved in collecting nearshore current data during hurricanes and other large storms make it difficult to verify Eq. 4.28. However, results from Eq. 4.28 are in general agreement with the rule of thumb that surface currents will be 2 or 3 percent of the wind speed. Eq. 4.28 can be used to calculate the current velocity field in the nearshore zone for a given wind field. From this, ΔS_c can be estimated by Eq. 4.27 using a step-type calculation similar to that used to calculate wind and bottom stress setup.

4.11. BATHYSTROPHIC APPROACH

With the estimated initial setup and the calculated pressure, wind and bottom stresses, and Coriolis setups for each of a sequence of offshore storm positions, the total storm surge hydrograph can be developed for a given coastal location by the procedures outlined in Section 4.10. An improvement on these procedures that entails a computer-applicable, numerical, finite difference approach is the bathystrophic storm-tide theory first proposed by Freeman, Baer, and Jung (1957).

The vertically integrated horizontal equations of continuity and motion (x-axis normal to shore, y-axis parallel to shore) neglecting vertical components of motion can be written

$$\frac{\partial S}{\partial t} + \frac{\partial q_x}{\partial x} + \frac{\partial q_y}{\partial y} = P \qquad (4.29)$$

$$\frac{dq_x}{dt} = -fq_y - gd\frac{\partial S}{\partial x} + \frac{\tau_{sx} - \tau_{bx}}{\rho} + gd\frac{\partial S_p}{\partial y} \qquad (4.30)$$

$$\quad\;\; \text{(a)} \quad\;\; \text{(b)} \quad\;\;\; \text{(c)} \quad\;\;\;\;\; \text{(d)} \quad\;\;\;\; \text{(e)}$$

$$\frac{dq_y}{dt} = -fq_x - gd\frac{\partial S}{\partial y} + \frac{\tau_{sy} - \tau_{by}}{\rho} + gd\frac{\partial S_p}{\partial y} \qquad (4.31)$$

In these equations

$$S = \text{surge elevation above SWL}$$

$$q_x = \text{discharge per unit width along } x\text{-axis}$$

$$q_y = \text{discharge per unit width along } y\text{-axis}$$

$$P = \text{precipitation rate}$$

$$f = \text{Coriolis parameter} = 2\omega \sin \phi$$

$$\tau_{sx}, \tau_{sy} = \text{wind stress along } x\text{-axis and } y\text{-axis}$$

$$\tau_{bx}, \tau_{by} = \text{bottom stress along } x\text{-axis and } y\text{-axis}$$

$$S_p = \text{atmospheric pressure head deficit}$$

The terms in the equations of motion (Eqs. 4.30 and 4.31) represent (a) total acceleration (convective plus local), (b) Coriolis force per unit mass, (c) hydrostatic gradient due to the water surface slope, (d) wind and bottom stress, and (e) atmospheric pressure effect.

The bathystrophic approach assumes:

1. The precipitation rate is neglectable ($P=0$).

2. The flow toward shore is neglectable ($q_x = 0$); thus Eq. 4.29 is not applied.

3. The convective acceleration is neglectable; thus

$$\frac{dq_y}{dt} = \frac{\partial q_y}{\partial t}.$$

4. The instantaneous sea surface elevation is constant parallel to the shore; thus $\dfrac{\partial s}{\partial y} = 0$.

Because of assumption 2, the x-axis, or computational axis, must be oriented essentially normal to the bed contours. Also, since $q_x = 0$, $dq_x/dt = 0$, which limits the approach to slow moving hurricanes and negligible flooding of dry land. Assumptions 2, 3, and 4 limit the approach to straight shorelines. The initial setup, wind-wave setup, and astronomical tide are determined separately and, if the pressure tide is also calculated separately,

Eqs. 4.30 and 4.31 become

$$gd\frac{\partial S}{\partial x} = -fq_y + \frac{\tau_{sx}}{\rho}$$ (4.32)

$$\frac{\partial q_y}{\partial t} = \frac{\tau_{sy} - \tau_{by}}{\rho}$$ (4.33)

The bottom stress term dropped out of Eq. 4.32 because of the assumption that $q_x = 0$. Allowance for bottom stress could be made, however, by combining it with the wind stress as done in Section 4.10e.

Relationships for the remaining bottom and wind stresses are inserted into Eqs. 4.32 and 4.33, which are then written in finite difference form. The equations are sequentially solved for q_y and S for small increments of time and space. The time and space increments may be nonuniform to allow for changes in rate of water level rise as a storm approaches shore and for irregularities in hydrography. Details of the finite difference forms of these equations, techniques for solving these equations, and example solutions (computer and manual) are given by Bodine (1971), U.S. Army Corps of Engineers (1973), and Pararas-Carayannis (1975). Marinos and Woodward (1968) present the results of bathystrophic storm surge calculations for Standard Project Hurricanes and other real hurricanes approaching shore at several sites along the Texas-Louisiana coast. For a large radius-moderate speed SPH approaching Sabine, Texas, the elevation and time of occurrence of the individual components and the total surge were computed to be:

Initial setup	2.0 ft (constant)
Wind stress setup	11.1 ft (38 hr)
Coriolis setup	2.6 ft (27 hr)
Pressure setup	1.8 ft (38 hr)
Total setup	14.1 ft (38 hr)

The above figures do not include the astronomical tide and wind wave setup.

4.12. TWO-DIMENSIONAL NUMERICAL MODELS

The previous methods for storm surge calculation are essentially one-dimensional, as calculations are carried out along a line normal to shore or across a bay. For irregular bays, estuaries and shorelines where continuity must be satisfied to obtain reasonable results and where flooding of low-lying terrain occurs, a more sophisticated method for storm surge calculations is usually needed. In recent years this need has been largely satisfied by the development of two-dimensional, vertically-integrated, finite difference numerical models. This development has depended heavily on the availability of high speed, large memory computers.

If important, pressure effects are calculated separately (Eq. 4.22), so the equations of motion can be written

$$\frac{\partial q_x}{\partial t} + \frac{q_x}{d}\frac{\partial q_x}{\partial x} + \frac{q_y}{d}\frac{\partial q_x}{\partial y} = -fq_y - gd\frac{\partial S}{\partial x} + kUU_x - Cqq_x \qquad (4.34)$$

$$\frac{\partial q_y}{\partial t} + \frac{q_x}{d}\frac{\partial q_y}{\partial x} + \frac{q_y}{d}\frac{\partial q_y}{\partial y} = -fq_x - gd\frac{\partial S}{\partial y} + kUU_y - Cqq_y \qquad (4.35)$$

where C is a bottom stress coefficient and $q = \sqrt{q_x^2 + q_y^2}$. The second and third terms on the left hand side are the convective acceleration or advection of momentum terms.

The area to be investigated is divided into square segments in a horizontal plane and an average depth and bottom stress coefficient are assigned to each square. Equations 4.29, 4.34, and 4.35 are written in finite difference form for application to each square segment in the mesh. At the sides of squares that act as boundaries the appropriate boundary conditions are written to allow for correct computation of flow behavior at these boundaries. Typical boundary conditions include: water-land boundaries through which there will never be any flow, adjacent low-lying squares that will flood when the water level reaches a certain elevation, inflow from rivers and surface runoff, barriers such as seawalls and dredge spoil dikes that can be overtopped, and offshore boundaries having a proscribed water-level time-history (e.g. due to astronomical tide).

The computer solution for the unknowns at each square segment (q_x, q_y, and S) proceeds over each row of segments in the mesh for given boundary conditions. Time is advanced and the spatial calculations repeated. The result is q_x and q_y at each side of every square segment and S for each segment, all at every time increment, although only selected values are printed.

It is imperative to have a good set of wind records and water level measurements from a past storm to calibrate the model. Calibration usually includes adjustment of the bottom stress coefficients and weir discharge coefficients for barriers, and so on. The precision achieved (and cost) depend to a great extent on the time interval and segment size selected. These also affect the mathematical stability of the model.

Ried and Bodine (1968) used such a two-dimensional numerical model to calculate storm surge elevations in Galveston Bay, Texas. Their model used Eqs. 4.29, 4.34, and 4.35 but neglected the convective acceleration and Coriolis terms. Flooding of low-lying terrain and overtopping of barriers were considered through application of the boundary conditions. Bottom friction was initially calibrated for a spring astronomical tide propagating into the bay during a period of calm wind conditions. Further calibration of bottom friction and discharge coefficients was done for the well-documented Hurricane Carla of 1961. The area studied is nominally 40 by 50 miles, which was subdivided into 2 mile square grids. Time steps of 3 and 4 min were used.

These two-dimensional and related numerical models are also used for a variety of other purposes including prediction of water quality, temperature and salinity variation, and tide levels (see Tracor, Inc., 1971).

4.13. CLIMATOLOGIC AND GEOLOGIC EFFECTS

Climatologic and geologic changes in sea level are often difficult to separate and usually occur at rates not significant to coastal engineering planning and design. However, there are instances of each that can be important such as the uplift and subsidence that occurred during the 1964 Alaska earthquake (Section 4.4).

Fairbridge (1961) presents a thorough discussion of sea-level fluctuations since the last glacial age and Hicks and Shofnos (1965) summarize sea-level changes around the United States during the twentieth century. At most locations, sea level is currently rising with average rates (1940–1962) of 0.34 cm/yr (North Atlantic); 0.27 cm/yr (South Atlantic); 0.21 cm/yr (Gulf); 0.06 cm/yr (Pacific); and a fall of -0.70 cm/yr (Southeast Alaska). Extreme local values include Eugene Island, Louisiana, 1.01 cm/yr, and Juneau, Alaska, -1.46 cm/yr. Neglecting shore line response, the horizontal position of the shoreline changes by the product of the vertical change times the cotangent of the bottom slope. Bruun (1962) discusses beach profile response to rising sea level.

While ocean level changes are due to long-term climatological variations superimposed on local geologic uplifting or subsidence, the larger water-level changes in the Great Lakes are primarily due to changes in seasonal and annual rainfall patterns. The lakes can experience mean monthly seasonal fluctuations of 0.3 m or more and longer term fluctuations of 1 to 2 m over a period of several years. High water levels cause serious shoreline erosion problems while low water levels cause navigation problems.

4.14. SUMMARY

This chapter presented the characteristics and, where possible, the techniques for predicting water level fluctuations having a period greater than the period of waves in the wind wave portion of the spectrum. Chapter 5 considers the generation, measurement, analysis, and prediction of wind-generated water waves.

4.15. REFERENCES

Abraham, G. (1961), "Hurricane Storm Surge Considered as a Resonance Phenomenon," *Proceedings, Seventh Conference on Coastal Engineering*, Council on Wave Research, University of California, Berkeley, pp. 585–602.

Biesel, F. (1955), "The Similitude of Scale Models for the Study of Seiches in Harbours," *Proceedings, Fifth Conference on Coastal Engineering*, Council on Wave Research, Grenoble, France, pp. 95–118.

Bodine, B. R. (1969), "Hurricane Surge Frequency Estimated for the Gulf Coast of Texas," Tech. Memo. 26, U.S. Army Coastal Engineering Research Center, 32 p.

Bodine, B. R. (1971), "Storm Surge on the Open Coast: Fundamentals and Simplified Prediction," Tech. Memo. 35, U.S. Army Coastal Engineering Research Center, 55 p.

Bretschneider, C. L. (1966), "Engineering Aspects of Hurricane Surge," in *Estuary and Coastline Hydrodynamics*, (A. T. Ippen, ed.) McGraw-Hill, New York, pp. 231–256.

Bretschneider, C. L. (1967), "Storm Surges," Vol. 4 *Advances in Hydroscience*, Academic Press, New York, pp. 341–418.

Bruun, P. (1962), "Sea-Level Rise as a Cause of Shore Erosion," *Journal, Waterways and Harbors Division*, American Society of Civil Engineers, February, pp. 117–130.

Carrier, G. F., Shaw, R. P. and M. Miyata (1971) "Channel Effects in Harbor Resonance," *Journal, Engineering Mechanics Division*, American Society of Civil Engineers, Dec., pp. 1703–1716.

Chow, V. T. (1959), *Open-Channel Hydraulics*, McGraw-Hill, New York, 680 p.

Cox, D. C. (1964), "Tsunami Height-Frequency Relationship at Hilo, Hawaii," Institute of Geophysics Report, University of Hawaii.

Cross, R. H. (1967), "Tsunami Surge Forces," *Journal, Waterways and Harbors Division*, American Society of Civil Engineers, Nov., pp. 201–231.

Defant, A. (1961), *Physical Oceanography*, Vol. II, Pergamon Press, New York, 598 p.

Dunn, G. E. and B. I. Miller (1964), *Atlantic Hurricanes*, Louisiana State University Press, Baton Rouge, 377 p.

Ewing, M. F., F. Press and W. L. Donn (1954) "An Explanation of the Lake Michigan Surge of 26 June, 1954," *Science* pp. 684–686.

Fairbridge, R. W. (1961), "Eustatic Changes in Sea Level," *Physics and Chemistry of the Earth*, Pergamon Press, Vol. 4, pp. 99–185.

Freeman, J. C., L. Baer and C. H. Jung (1957), "The Bathystrophic Storm Tide," *Journal of Marine Research*, Vol. 16, pp. 12–22.

Gilmour, A. E. (1961) "Tsunami Warning Charts," *New Zealand Journal of Geology and Geophysics*, Vol. 4, p. 132–5.

Graham, H. E. and D. E. Nunn (1959), "Meteorological Considerations Pertinent to Standard Project Hurricane, Atlantic and Gulf Coasts of the United States," National Hurricane Research Project Rept. 33, U.S. Dept. of Commerce, 76 p.

Hicks, S. D. and W. Shofnos (1965), "Yearly Sea Level Variations for the United States," *Journal, Hydraulics Division*, American Society of Civil Engineers, September, pp. 23–32.

Iida, K. (1969), "The Generation of Tsunamis and the Focal Mechanism of Earthquakes," Chap. 1., *Tsunamis in the Pacific Ocean* (W. M. Adams, ed.), East-West Center Press, U. of Hawaii, pp. 3–18.

Kamel, A. (1970), "Laboratory Study for Design of Tsunami Barrier," *Journal, Waterways, Harbors and Coastal Engineering Division*, American Society of Civil Engineers, Nov., pp. 767–779.

Kaplan, K. (1955), "Generalized Laboratory Study of Tsunami Run-up," Tech. Memo. 60, U.S. Army Beach Erosion Board, Washington, D.C., 30 p.

Keulegan, G. H. and J. Harrison (1970), "Tsunami Refraction Diagrams by Digital Computer," *Journal, Waterways and Harbors Division*, American Society of Civil Engineers, May, pp. 219–233.

Macmillan, D. H. (1966), *Tides*, American Elsevier, New York, 240 p.

Magoon, O. T. (1965), "Structural Damage by Tsunamis," *Proceedings, Santa Barbara Coastal Engineering Specialty Conference*, American Society of Civil Engineers, pp. 35–68.

Marinos, G. and J. W. Woodward (1968), "Estimation of Hurricane Surge Hydrographs," *Journal, Waterways and Harbors Division*, American Society of Civil Engineers, May, pp. 189–216.

Miles, J. W. (1948), "Coupling of a Cylindrical Tube to a Half Space," *Journal, Acoustic Society of America*, pp. 652–664.

Miles, J. W. (1974), "Harbor Seiching," *Annual Review of Fluid Mechanics*, Vol. 6, pp. 17–35.

Myers, V. A. (1954), "Characteristics of U.S. Hurricanes Pertinent to Levee Design for Lake Okeechobee, Florida," H.R. Report 32, U.S. Weather Bureau.

Neumann, G. and W. J. Pierson (1966), *Principles of Physical Oceanography*, Prentice-Hall, Englewood Cliffs, New Jersey, 545 p.

Pararas-Carayannis, G. (1975), "Verification Study of a Bathystrophic Storm Surge Model," Tech. Memo. 50, U.S. Army Coastal Engineering Research Center, 248 p.

Peterson, K. R. and H. V. Goodyear (1964), "Criteria for a Standard Project Northeaster for New England North of Cape Cod," National Hurricane Research Project Rept. 68, U.S. Dept. of Commerce, 66 p.

Raichlen, F. (1966), "Harbor Resonance," in *Estuary and Coastline Hydrodynamics*, (A. T. Ippen, ed.), McGraw-Hill, New York, pp. 281–340.

Ried, R. O. and R. B. Bodine (1968), "Numerical Model for Storm Surges in Galveston Bay," *Journal, Waterways and Harbors Division*, American Society of Civil Engineers, February, pp. 33–57.

Roberts, E. B. (1950), "A Seismic Sea Wave Warning System for the Pacific," *U.S. Coast and Geodetic Survey Journal*, No. 3, pp. 74–79.

Saville, T. (1952), "Wind Set-Up and Waves in Shallow Water," Tech. Memo. 27, U.S. Army Beach Erosion Board, Washington, D.C., 36 p.

Schureman, P. (1940), *Manual of Harmonic Analysis and Prediction of Tides*, Special Publication 98, Coast and Geodetic Survey, Dept. of Commerce, Washington, D.C., 317 p. (reprinted with corrections 1958).

Sorensen, R. M. and W. N. Seelig (1976), "Hydraulics of Great Lakes Inlet-Harbor Systems," *Proceedings, Fifteenth Conference on Coastal Engineering*, American Society of Civil Engineers, Honolulu, 20 p.

Spaeth, M. G. and S. C. Berkman (1965) "The Tsunami of March 28, 1964 as Recorded at Tide Stations," U.S. Coast and Geodetic Survey Report, 59 p.

Tracor, Inc. (1971), "Estuarine Modeling: An Assessment," Report for Water Quality Office, Environmental Protection Agency, U.S. Government Printing Office, 497 p.

U.S. Army Coastal Engineering Research Center (1973), *Shore Protection Manual*, 3 Vols., U.S. Government Printing Office, Washington, D.C.

U.S. Army Engineer District, New Orleans (1970), "Report of Hurricane Camille, 14–22 August 1969."

U.S. Weather Bureau (1968), "Interim Report-Meteorological Characteristics of the Probably Maximum Hurricane, Atlantic and Gulf Coasts of the United States," HUR-7-97, U.S. Dept. of Commerce.

Van Dorn, W. G. (1965), "Tsunamis," Vol. 2, *Advances in Hydroscience*, Academic Press, New York, pp. 1–48.

Van Dorn, W. G. (1953), "Wind Stress on an Artificial Pond," *Journal of Marine Research*, Vol. 12, pp. 249–276.

Wilson, B. W. (1960), "Note on Surface Wind Stress over Water at Low and High Wind Speeds," *Journal of Geophysical Research*, Vol. 65, pp. 3377–3382.

Wilson, B. W. (1962), discussion of "Harbor Paradox," by J. Miles and W. Munk, *Journal, Waterways and Harbors Division*, American Society of Civil Engineers, May, pp. 185–194.

Wilson, B. W. (1972), "Seiches" Vol. 8, *Advances in Hydroscience*, Academic Press, New York, pp. 1–94.

Wilson, B. W. and A. Tørum (1968), "The Tsunami of the Alaskan Earthquake, 1964: Engineering Evaluation," Tech. Memo. 25, U.S. Army Coastal Engineering Research Center, 445 p.

Wiegel, R. L. (1970) "Tsunamis," Chap. 11, *Earthquake Engineering* (Wiegel, ed.), Prentice-Hall, Englewood Cliffs, New Jersey, pp. 253–306.

Zetler, B. D. (1947), "Travel time of Seismic Sea Waves to Honolulu," *Pacific Science*, Vol. 1, pp. 185–8.

4.16. PROBLEMS

- 1. Using the National Ocean Survey Tide Tables, plot the tide curve at Ashbury Park, New Jersey, for July 4*th* of this year. Is it closer to being a spring or a neap tide?
- 2. Over a great circle route between an earthquake epicenter on the coast of Chile and Hawaii the average depth is 50 fathoms for 150 n.m., 1800 fathoms for 390 n.m., 2200 fathoms for 3870 n.m., and 2800 fathoms for

1260 n.m. Calculate the tsunami travel time between these points assuming a path along the great circle route.

· 3. A tsunami wave has a 0.7 m height in water 2000 m deep. If refraction causes orthogonal spacing to decrease by a factor of two at a depth of 20 m, what is the wave energy per foot of crest width at the 20 m depth?

4. Calculate the fundamental periods for longitudinal and lateral oscillations in the Coastal Engineering Research Center wave tank considered in Prob. 2.1. Assume a water depth of 4 m. Does it appear that the fundamental or harmonic modes of the longitudinal or lateral oscillation will present any difficulties to experimenters?

5. A pair of two-dimensional basins have the geometries shown. Calculate the fundamental period of free oscillation for each.

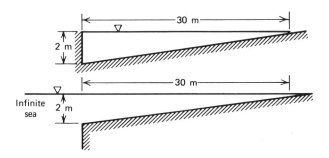

6. A bay has the horizontal dimensions and bottom contours shown. Determine the fundamental period of longitudinal free oscillation. If the jettied entrance is 200 m long, 30 m wide and 5 m deep, calculate the Helmholtz period.

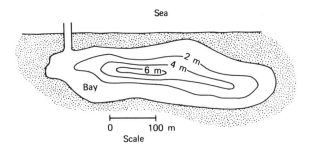

7. Calculate the fundamental and first two harmonic periods of longitudinal oscillation for Lake Michigan.

8. A 3 m deep, rectangular, small boat harbor has horizontal dimensions of 200 m by 400 m. What are its four longest periods of free oscillation? For $H=0.5$ m at the $n=1$, $m=1$ mode, what is the largest horizontal velocity developed in the harbor?

9. For Prob. 8 with $n=2$, $m=3$ and $H=0.4$ m, sketch and label water surface contours at $T/3$. What is the water level range and maximum velocity at $x=y=50$m?

10. Explain the development of Eq. 4.18 from basic Froude number similarity.

11. A 50 knot wind blows in a direction parallel to the axis of the bay shown in Prob. 6. Determine the resulting water level difference between the leeward and windward ends of the bay.

12. Assume that the nodal point is at the midpoint of the axis in prob. 11 and calculate the resulting bay water surface profile along the axis. Check the resulting profile for water volume balance.

13. A large radius SPH is approaching shore with the point of highest winds approaching Galveston, Texas, along a line S 10° E. When the point of highest winds is 30 n.m. offshore, calculate the pressure, and wind and bottom stress setups at Galveston. Estimate the Coriolis setup.

14. The bottom profile off a section of the Mississippi coast has the following depths at the given distances from shore: 2.7 m (3 n.m.), 3.8 m (7 n.m), 4 m (10 n.m.), 11 m (20 n.m.), 12.2 m (30 n.m.), 15.5 m (40 n.m.), 27 m (50 n.m.), 48 m (60 n.m.), 73 m (70 n.m.), and 200 m (80 n.m.). With the SPH ($R=26$ n.m.) from Fig. 4.11 located so the point of maximum winds is 25 n.m. from shore, calculate the pressure setup and the wind and bottom stress setup on shore. Estimate the Coriolis setup and total setup if the initial setup is 0.6 m.

PLOT TIDES FOR WEEK OF 15 FEB, 77

FOR ANNAPOLIS

WIND-GENERATED WAVES

The most apparent and often most important waves in the spectrum of waves at sea are those generated by the wind. Figure 5.1 shows a typical water-surface time-history recorded at a stationary location at sea and demonstrates the irregular nature of wind-generated waves. A wave record taken at the same time at a nearby location would probably look quite different but would have very similar statistical properties. The wave record for a particular location may often contain locally generated wind waves superimposed on typically longer and lower swell generated at some distant location (plus the tide and other long period components). It is also common for a train of wind-generated waves to consist of alternating groups of high and low waves similar to the pattern of waves shown in Fig. 2.7, but with greater irregularity.

As the wind velocity, distance or fetch over which the wind blows, and/or duration of the wind increase, the average height and period of the wind-generated waves will increase (within limits). For a given wind speed and unlimited wind fetch and duration there is a fixed limit to which the average wave height, period, and spectral energy will grow. At this limiting condition, the rate of energy input from the wind to the waves is balanced by the rate of wave energy dissipation due to wave breaking and turbulence. This condition, known as a fully developed sea, is used for the development of many standardized wind wave spectra.

When waves leave the generating area, their surface profile becomes smoother and more symmetrical and the average height of the waves begins to decrease due to energy dissipation by air resistance and turbulence,

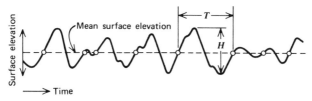

Figure 5.1   Water-surface time-history record.

frequency dispersion, and lateral spreading. Since wave celerity depends on the period, frequency dispersion causes a wave group to spread (in the travel direction) with time as the longer waves move ahead and shorter waves fall behind.

McClenan and Harris (1975) conducted a study of aerial photographs of swell at sea and in the nearshore zone. They found a high incidence of one or more distinct wave trains having long crests and fairly constant wave periods. In several photographs they could identify as many as four or five distinct wave trains coming from different directions. In the offshore region the shorter and steeper waves were usually most visible. However, the predominant surf zone waves in the same photo were the longer swell, not as discernable offshore, that undergo the greatest increase in steepness while shoaling. These longer waves tend to dominate the surf zone hydraulic and sediment-transport processes. At some nearshore locations the large number of wave trains observed is due partially to reflection and refraction-diffraction effects that cause a wave train to overlap itself (e.g., see Fig. 3.7b).

Examination of a typical wave record (Fig. 5.1) raises questions such as which undulations of the water surface should be considered as waves and what are the individual heights and periods of these waves? The approach used to analyze a wave record must be reasonable and consistent. Of the approaches used, the most common is the zero-upcrossing method (Pierson, 1954). A mean water surface elevation is drawn on the record and each point where the water surface crosses the mean water level in the upward direction is marked (circles on Fig. 5.1). The time elapsed between consecutive points is a wave period and the maximum vertical distance between crest and trough is the wave height.

It is often desirable to select a single wave height and period to represent a spectrum of waves for use in wave prediction, wave climate analysis, design of coastal structures, and so on. If the wave heights from a representative record are ordered by size one can define a height H_n that is the average of the upper n percent of the wave heights. For example, H_{10} is the average height of the highest 10 percent of the waves in the record. The most commonly used representative wave is H_{33} or the average height of the upper third of the waves. This is called the significant wave height H_s and it is approximately the height an experienced observer will give when visually estimating the height of waves at sea. The same approach can be applied to wave period but the most common representative period (called the significant period T_s) is usually defined as the average period for the highest one-third of the waves, as the period for the peak of the energy spectrum, or just as the average period for the record. Usually, these three definitions will produce about the same period unless, for example, the energy spectrum is bimodal due to the existence of wind waves and swell.

5.1. WIND WAVE GENERATION

The characteristics of a spectrum of wind-generated waves depend primarily on the fetch length F, wind velocity U, wind duration t_d, and to a lesser, but often significant, extent on the fetch width, water depth, bottom friction characteristics, atmospheric stability, and spatial and temporal variations in the wind field during wave generation.

Waves are generated with travel directions aligned at a range of oblique angles $(<90°)$ with the direction of the wind. The range of directions decreases with an increase in wave period as waves grow while propagating along a fetch. Thus the smaller the fetch width the less chance shorter waves have of remaining in the generating area and growing to significant size. The water depth affects the wave surface form and kinematics and thus the transfer of energy from wind to waves. Water depth also limits the nonbreaking wave heights. Of course, bottom friction dissipates wave energy and thus decreases both the rate at which waves will grow and the ultimate wave size. Atmospheric stability, which depends on the air/sea temperature ratio, affects the wind velocity profile near the sea surface, and the related wind-shear stress and form drag on the water surface.

It is instructive to consider the variation of significant wave height and period as a function of generating distance for waves generated by a wind of constant velocity, blowing over a constant fetch, and having different durations. This is demonstrated schematically in Fig. 5.2. If the wind duration exceeds the time required for waves to travel the entire fetch length (i.e. $t_d > F/C_g$) the waves will grow to OAB and their characteristics at the end of the fetch will depend on F (for given U). This is the "fetch limited" condition. If the duration is less (i.e. $t_d < F/C_g$) wave growth stops at OAC $(x = F_{min})$ and wave generation is "duration limited." Wave growth with

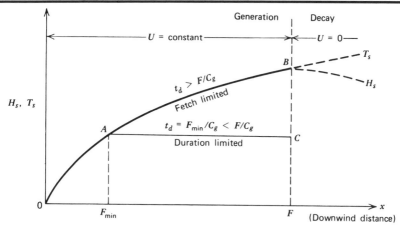

Figure 5.2   Schematic of wave generation for constant wind velocity.

time is given by curves shaped like OAC that stop at OAC or OAB depending on whether conditions are duration or fetch limited. If both the fetch and duration are sufficiently large the curve OAB becomes essentially horizontal at the downwind end and a fully developed sea (FDS) has been generated for that particular wind velocity. In the decay region, beyond the end of the fetch, waves propagate as swell and the significant height decreases while the significant period increases as indicated in Fig. 5.2.

In Section 2.4 it was shown that the wave energy per unit surface area for a monochromatic wave is proportional to the wave height squared. A record of the irregular sea-surface time-history at a point consists of many component waves having a range of periods and heights. If a short record is analyzed to determine the height and period of all existing components and the sums of the squared heights for the waves in each of several period or frequency intervals are determined, an energy-period or energy-frequency spectrum can be plotted. The former would have typical ordinates of $m^2/\sec$ versus period in seconds and the latter m^2-sec versus frequency in $\sec^{-1}$ or hertz. Occasionally, wave energy-frequency spectra are plotted as a function of wave angular frequency, $2\pi/T$ in rad/sec. Wave energy spectra are very useful for certain design analyses and they define certain important wave characteristics. However, they mask some information such as the sequence and phasing of high and low waves and the correlation between height and period for individual waves.

Figure 5.3 shows the growth of typical period spectra for points along a fetch up to development of the FDS for a constant wind velocity. The FDS spectrum would not change for fetches larger than that necessary to generate

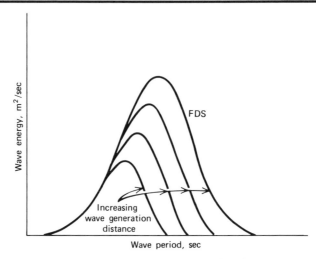

Figure 5.3 Typical wind wave spectra at points along a fetch for constant wind velocity.

the FDS. The shorter period waves grow to an energy level limited by breaking and remain near this level throughout the remainder of the fetch. The range of wave periods over which this limit applies grows with distance along the fetch as the spectrum grows. Since wave energy transfer is proportional to H^2T, longer period waves require a greater energy input for a given increase in height. The result is that the highest waves and greatest energy concentration are at or near the average period, and the energy-period spectrum curves are approximately symmetrical. Thus the earlier stated conclusion that the average period, the average period of the one-third highest waves, and the period of peak energy are often approximately equal.

If spectra were evaluated for wave records taken at a fixed point in a wave generating area at successive times, the curves would exhibit the same growth with time as shown in Fig. 5.3 for increasing generation distances. The final spectra curve would be the FDS if the fixed measuring point was at the downwind end of the fetch.

The actual mechanisms by which wind waves are generated are not completely understood but significant insight has been provided by theories proposed and experimentally evaluated by certain investigators. Some of the wind wave-generation theories provide the basis (with calibration using empirical data) for existing wave prediction techniques. Highlights will be summarized chronologically in the remainder of this section.

Jeffreys (1925), assuming the existence of a wave form and neglecting tangential wind stress on the water surface, evaluated the form drag due to normal wind pressure variation along the wave profile. This depends on the wave amplitude, profile shape, celerity, the wind velocity, and a "sheltering coefficient" that relates the surface pressure variation to the profile shape. Equating energy input to the wave through wind pressure to viscous energy dissipation within the wave (turbulent dissipation was ignored), Jeffreys obtained a relationship for wave growth due to the wind. His results suggested wave growth would occur only when the wind speed exceeded the wave celerity, and a minimum wind speed of 1.1 m/sec is needed to generate waves.

The tangential wind stress on the water surface, which depends on the difference between the wind velocity and the water particle velocity at the surface, is also an important wave generation mechanism. It often makes a greater contribution to wave generation than normal pressure. Since the surface particle velocity is less than the wave celerity (except near breaking) it is possible for the wind to transfer energy to waves through tangential stress even when the wave celerity exceeds the wind speed. Sverdrup and Munk (1947) considered both the normal pressure and tangential stress energy transfer mechanisms to develop a theory for wave generation. When the "wave age" defined as $C/U < 0.37$ they found the pressure mechanism to predominate, but for $C/U > 0.37$ energy transfer was predominantly through tangential stress. Wave growth continues until $C/U = 1.37$ when the FDS occurs. To apply empirical data to evaluate their theory for wave generation they evolved the concept of significant wave height and significant period to represent a wave record.

Neither Jeffreys nor Sverdrup and Munk dealt with the initial generation of waves on a smooth water surface as wind velocity increases from zero. A wind field contains a range of small- to large-scale three-dimensional eddies that exert random fluctuations of pressure and shear on the water surface as they propagate forward at approximately the average wind speed U. The size distribution, internal velocity and pressure fluctuations, duration, and forward speed of these eddies depend on the wind speed, surface roughness and atmospheric stability. It is apparent that the random pressure and tangential stress fluctuations in these eddies can distrub a smooth surface and initiate the generation of small waves.

Phillips (1957) developed a wave generation theory by considering only the pressure fluctuations in a turbulent wind field. Vital to this theory is the resonant interaction between wind eddies and selected waves having the same forward speed. Thus the growth rate is maximum for those waves in the spectrum having $C/U \cong 1$, but growth can occur at $C/U \gtrsim 1$. Through the three-dimensional characteristics of turbulent fluctuations, Phillips was able to explain the generation of waves with travel directions oblique to the direction of the wind.

Miles (1957) introduced the shear-flow mechanism concept for wind wave generation. This involves the secondary circulation set up around an axis parallel to the wave crest by a logarithmic wind-velocity profile acting over a moving sinusoidal wave surface profile. Below a point in the wind velocity profile where the velocity equals the wave celerity, air flow is backward relative to the wave celerity. Above this point, air flow is in the direction of wave motion. This results in a circulation cell moving at the wave celerity and exerting a pressure gradient on the wave surface that transfers energy to the waves. The energy transfer rate is proportional to the square of the wave height so steeper waves are selectively amplified. The concept neglects flow separation in the lee of the wave crest and is thus valid only for lower amplitude waves.

In summary, it appears that all the mechanisms discussed (and others yet to be developed) contribute to the generation of wind waves. Initially, turbulent fluctuations disturb the surface to develop waves and contribute to their growth. As waves grow the shear-flow mechanism assumes increasing importance, particularly for shorter period waves. When air flow separates from the wave profile at the crest, form drag increases in importance. Throughout most of the wave growth, tangential wind stresses are important. Of course, maximum growth is limited by the wind energy available, the time available for energy transfer, and wave instability and breaking. Another mechanism that is involved to some degree is the transfer of energy from shorter to longer period waves, when shorter waves break and when nonlinear interactions occur between shorter and longer waves.

5.2. WAVE STATISTICS

Because of their random nature the characteristics of wind waves are most easily handled statistically. Two common types of wind wave record analysis are: 1) evaluation of the probability distribution of wave heights in order to determine the mean wave height, H_{33}, H_{10}, and so on, and 2) determination of the wave energy spectrum. Also of importance to engineering design is the determination of the return period or expected occurrence frequency of a particular extreme wave condition (e.g., the significant height of the 50-year storm) at a site. These and related matters will be covered in this section.

Longuet-Higgins (1952) derived a relationship for the statistical distribution of wind wave heights assuming that the wave spectrum contains a single narrow band of frequencies and that wave energy comes from a number of sources whose phases are random. These assumptions required that the waves being analyzed be from a single storm a great distance away, so that frequency dispersion narrows the band of wave frequencies recorded. Bretschneider (1959) and Wiegel (1964) summarized many data sources that demonstrate the general applicability of Longuet-Higgins' results to wave records from both distant and nearby storms. Goodknight and Russell (1963) and Collins (1967) showed the applicability of the results to waves measured in and near hurricanes.

Longuet-Higgins demonstrated that the probability of occurrence $p(H_i)$ of a particular wave height H_i in a record containing N waves can be specified by a Rayleigh distribution

$$p(H_i) = \frac{2H_i}{(H_{rms})^2} e^{-(H_i/H_{rms})^2} \tag{5.1}$$

Here, the root mean square height H_{rms} is given by

$$H_{rms} = \sqrt{\frac{1}{N} \sum_i H_i^2} \tag{5.2}$$

The mean wave height H_{100} is then

$$H_{100} = \frac{\int_0^{\infty} H_i p(H_i)\, dH_i}{\underbrace{\int_0^{\infty} p(H_i)\, dH_i}_{\longrightarrow 1.0}} = \frac{\sqrt{\pi}\, H_{rms}}{2} \tag{5.3}$$

The cumulative probability distribution $P(H_i)$ [i.e., fraction of waves $\leqslant H_i$] is

$$P(H_i) = \int_0^{H_i} p(H_i)\, dH_i = 1 - e^{-(H_i/H_{rms})^2} \tag{5.4}$$

Eq. 5.4 leads to the following distribution of wave heights:

n	H_n/H_{33}	H_n/H_{100}
1	1.68	2.68
10	1.28	2.03
33	1.00	1.60
50	0.89	1.42
100	0.63	1.00

For example, the average height of the highest 1 percent of the waves can be expected to be 1.68 times the significant height and 2.68 times the average height.

Longuet-Higgins also showed that the maximum wave height, H_{max}, in N waves is

$$H_{max} = 0.707 H_{33} \sqrt{\ln N} \qquad (5.5)$$

Thus in a storm containing 5000 waves $H_{max} = 2.06 H_{33}$ and $H_{max} = 3.31 H_{100}$. Since $H_{100}/H_{rms} = \sqrt{\pi}/2$ and $H_{33}/H_{100} = 1.60$, $H_{33}/H_{rms} = 1.41$. From Eq. 5.4

$$P(H_{33}) = 1 - e^{-(1.41)^2} = 0.865$$

So $1 - P(H_{33}) = 0.135$, or 13.5 percent of the waves can be expected to exceed the significant height.

If the water surface deviation from the still water level as a function of time is η_t, the standard deviation σ of η_t is

$$\sigma = \sqrt{\frac{\sum \eta_t^2}{N^*}}$$

where N^* is the number of points along the wave record at which η_t is measured. Thus the variance

$$\sigma^2 = \frac{\sum \eta_t^2}{N^*} = \overline{\eta_t^2}$$

the average value of η_t^2. For a record length T^* the potential energy density $\overline{E}_p$ is

$$\overline{E}_p = \frac{1}{T^*} \int_0^{T^*} \rho g \eta_t \left(\frac{\eta_t}{2} \right) dt$$

$$= \frac{\rho g}{2} \overline{\eta_t^2} = \frac{\rho g \sigma^2}{2}$$

Assuming $\overline{E}_p = \overline{E}/2$

$$\overline{E} = \rho g \sigma^2 \qquad (5.6)$$

For a wave spectrum

$$\bar{E}=\frac{\rho g\left(\dfrac{\sum H_j^2}{N}\right)}{8}=\frac{\rho g\left(H_{rms}\right)^2}{8}$$

where H_j is the height of each of the N waves that combine to form the spectrum of waves. Since $H_{33}/H_{rms}=1.41$

$$\bar{E}=\frac{\rho g\left(H_{33}\right)^2}{16} \tag{5.7}$$

Equation 5.7 shows that the energy density of the significant height is twice the energy density of the wave spectrum (assuming the Rayleigh distribution applies). Also, from Eqs. 5.6 and 5.7

$$\rho g\sigma^2=\frac{\rho g}{16}\left(H_{33}\right)^2$$

or

$$H_{33}=4\sigma \tag{5.8}$$

Bretschneider (1959) demonstrated that the Rayleigh distribution can also be used to describe the distribution of wave lengths. Since, $L=gT^2/2\pi$ in deep water, this yields

$$p(T_i)=2.7\frac{(T_i)^3}{(\bar{T})^4}e^{-0.675(T_i/\bar{T})^4} \tag{5.9}$$

and

$$P(T_i)=1-e^{-0.675(T_i/\bar{T})^4} \tag{5.10}$$

where $\bar{T}$ is the mean wave period. Collins (1967) found that Eqs. 5.9 and 5.10 showed reasonable agreement with the distribution of higher period waves in Hurricane Dora that are of greatest engineering significance.

It is desirable to have a simple yet meaningful method for analyzing each of a large number of wave records to determine the significant wave height and period for each record. Many methods have been used but each will give slightly different results from the same wave record. This is undesirable when one tries to compare wave data from several sources.

The National Institute of Oceanography, Great Britain (Draper, 1966; Tucker, 1963) has a method to analyze pen and ink wave records. A mean water level is drawn by eye and all zero upcrossing points are marked. The duration of the record divided by the number of zero upcrossings N_z yields an average or significant zero upcrossing period T_z. Then the highest crest and lowest trough in the record (whether or not from the same wave) are measured and summed to yield a maximum wave height. A factor, which is based on the work of Longuet-Higgins and others and is a function of N_z, is multiplied times this maximum wave height to yield the significant height. This method can be programmed for a rapid computer analysis of wave records.

The pen and ink wave record analysis method currently used by the U.S. Army Coastal Engineering Research Center (typically on 7-min records) uses a transparent template to estimate the period of the higher and more uniform waves in the record. The template contains several series of uniformly spaced parallel lines with spacings scaled to represent typical wave periods. The record length divided by this significant period gives an estimate of the number of waves in the record. The number of waves multiplied by 0.135 gives the rank of the wave having the significant height. This wave is located and its crest to trough height recorded as the significant wave height.

Analysis of water-surface time-history records of, say, 15-20 min duration to obtain the ocean wave energy spectrum requires a computer analysis using very complex procedures. Discussions of these procedures are given by Blackman and Tukey (1958), Kinsman (1965), and Harris (1974). Spectral analysis procedures were improved and made easier to apply with the introduction of the Fast Fourier Transform (FFT) by Cooley and Tukey (see IEEE, 1967). The result is that wave energy spectra for a large number of wave records can now be determined on a routine basis.

Digital wave records recorded on magnetic tape by the U.S. Army Coastal Engineering Research Center are routinely analyzed (Thompson, 1974) to yield energy spectra (by FFT) as well as significant height and period. The significant period is the period having the maximum energy density from the spectrum and the significant height is determined from the standard deviation of the water surface elevation using Eq. 5.8.

Several authors have proposed wave energy spectrum models based on a combination of theoretical reasoning and empirical data. From previous discussions, the period spectrum

$$S_{H^2}(T) = f(T, g, U, F, t_d) \qquad (5.11)$$

where $S_{H^2}(T)$ is the summation of H^2 as a function of T.

Bretschneider (1959) proposed a spectrum based on the period and height probability distributions (Eqs. 5.1 and 5.9), assuming zero correlation between individual wave periods and heights. This may be written in terms of the mean period $\bar{T}$ and mean height $\bar{H}$ (i.e. H_{100}) as

$$S_{H^2}(T) = 3.44 \frac{(\bar{H})^2 T^3}{(\bar{T})^4} e^{-0.675(T/\bar{T})^4} \tag{5.12}$$

$\bar{H}$ and $\bar{T}$ depend, of course, on g, U, F, and t_d as discussed in Sections 5.1 and 5.5.

Pierson and Moskowitz (1964) analyzed many wave records taken by British weather ships in the North Atlantic. They selected records representing fully developed seas for wind speeds between 20–40 knots to establish the spectrum

$$S_{H^2}(T) = \frac{8.1 \times 10^{-3} g^2 T^3}{(2\pi)^4} e^{-0.74(gT/2\pi U)^4} \tag{5.13}$$

In Eq. 5.13, U represents the wind velocity at an elevation of 19.5 m above sea level, which is typically about 10 percent higher than the velocity at the standard elevation of 10 m (Bretschneider, 1965). The Pierson-Moskowitz spectrum has been widely used as a design spectrum but it must be remembered that it was developed for FDS conditions. Other spectra in use have been presented by Neuman (1952) and Hasselmann et al. (1973).

Integration of Eqs. 5.12 or 5.13 from $T=0$ to ∞ for the area under the curve yields the total specific energy (energy per unit area) divided by ρg. Thus

$$\bar{E} = \rho g \int_0^\infty S_{H^2}(T) dT \tag{5.14}$$

and, from Eq. 5.6, $\bar{E} = \rho g \sigma^2$, so the area under the spectrum curve equals the variance σ^2 and

$$\sigma^2 = \left(\frac{H_{33}}{4}\right)^2 \tag{5.15}$$

from Eq. 5.8. Integrating Eq. 5.13 from 0 to ∞ for T and applying Eq. 5.8 yields

$$H_{33} = 0.0056 U^2 \tag{5.16}$$

for the Pierson-Moskowitz spectrum with H_{33} in meters and U in knots. Differentiating Eq. 5.13 with respect to T and equating to zero yields the period for the peak of the spectrum, which is the significant or mean period $\overline{T}$. This operation yields

$$\overline{T} = 0.33 U \qquad (5.17)$$

with $\overline{T}$ in seconds and U in knots. Thus Eqs. 5.16 and 5.17 give the significant wave height and period for a FDS. For a 40-knot wind velocity measured at 19.5 m elevation and a FDS, Eqs. 5.16 and 5.17 yield $H_{33} = 8.95$ m and $\overline{T} = 13.2$ sec. The resulting Pierson-Moskowitz period spectrum (Eq. 5.13) is plotted in Fig. 5.4.

Besides the analysis of wave records to obtain T, H_{33}, H_{max}, and so on, it is often necessary for design or construction planning purposes to look at long-term variations in wave characteristics. For example, Thompson and Harris (1972) presented plots of significant height versus percent of time that

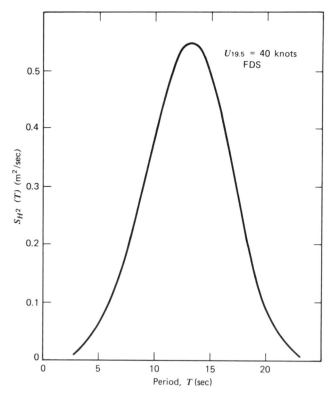

Figure 5.4 PM spectrum for 40-knot wind.

height was exceeded during a year at a particular location. The data are based on one 7-min record each day which was demonstrated to be sufficient. Similar analyses for 10 consecutive years at one site showed the annual variability in the height distributions to be small enough for 1 year of data to be sufficiently representative.

Yang et al. (1974) selected the annual maximum significant wave height from 24 years of shipboard wave observations offshore of Delaware. These data were fitted with a Gumbel type 1 distribution to develop a relationship for return period in years of a given maximum annual wave height. If sufficient wave data are not available but long-term wind records are, hindcasted waves (see Section 5.5) may be used for a return period analysis. Draper (1972) discusses a study to determine 50-year return period design wave characteristics for the British Isles from 50-year return period wind conditions.

5.3. SOURCES OF WIND WAVE INFORMATION

Wind wave data can be obtained from a variety of sources. Specific wave data requirements depend on the use to which the data will be put. A few design wave needs are discussed below in order to demonstrate the types and quantities of wave data needed for different uses.

FIXED RIGID STRUCTURES (E.G., OFFSHORE DRILLING TOWER)   Either a design wave height and period, or preferably, a design wave spectrum, are needed. Structure failure is likely to be catastrophic rather than gradual. A design wave might have the most probable maximum height having a selected return period (say 100 years). A range of reasonable wave periods might be used with this height to determine the critical conditions. Structure resonant periods would likely be critical. The design wave spectrum might be that giving the most probably critical loading on the structure for a sea state with a selected return period.

FLEXIBLE STRUCTURES (E.G., RUBBLE MOUND BREAKWATER)   As failure is gradual and structures are repairable, it is common to design for the significant rather than maximum wave height and for a lower return period than for rigid structures. For structures located nearshore, water depths may be shallow enough so the maximum wave height is limited by breaking.

BEACH PROCESSES (E.G., PREDICTION OF LONGSHORE SEDIMENT TRANSPORT RATES)   Study of beach processes ideally requires a knowledge of the annual percent exceedence of daily mean wave heights, periods, and nearshore directions as well as seasonal variations in these characteristics. Nearshore direction is of great importance in many cases and is difficult to accurately measure on a continuous basis.

SEAWALLS   The nearshore wave spectrum for the design return period is needed. Runup of the spectrum of waves on the seawall establishes the crest elevation and amount of flooding by overtopping of the crest. The significant height or perhaps a higher wave such as H_{10} would be used for an analysis of seawall stability characteristics depending on the structural characteristics of the seawall as well as its intended function.

COASTAL CONSTRUCTION PLANNING   The percent exceedence of a particular significant or other height on an annual, seasonal, or monthly basis is needed for an economic analysis and planning of work schedules. On a day-by-day basis, for example, when an expensive piece of equipment with limited seakeeping ability is to be used, accurate wave forecasts for the next few days are often vital.

The remainder of this section is a brief survey of some of the primary types of wave data available to the practicing coastal engineer.

Wave Gaging Programs

Programs to measure wave characteristics at selected coastal locations have been conducted by universities, a few federal agencies, and private corporations, particularly the major oil companies. These programs usually involve the installation of a wave gage to record the water-surface time-history at a point for a period of, say, 20 min every 4 or 6 hr for at least a year. The data are recorded locally on a strip chart or preferably on magnetic tape, or are transmitted back to a central location for recording. Although wave direction is very important, no completely successful device or procedure has been developed to date for the continuous measurement of wave direction. Multiple wave gage arrays, radar, air photography, and other devices have seen limited use for this purpose.

The common types of wave gages are discussed below. See Grace (1970), Peacock (1974), and Ribe and Russin (1974) for additional information and references on the large variety of wave gages available.

PRESSURE GAGE   As demonstrated in Section 2.3, a device that measures underwater dynamic pressure fluctuations can be used to determine wave heights and periods, through application of Eq. 2.32. Owing to wave pressure decay with increasing water depth, a pressure gage filters the higher frequency end of a wave spectrum, the degrees of filtering depending on the depth at which the pressure sensor is placed and the range of periods in the wave spectrum. Data are stored underwater with the sensor or transmitted back to shore via a cable.

STEP-RESISTANCE GAGE   A vertical staff with electrodes spaced at a discrete interval of 0.1 or 0.2 ft pierces the water surface. As the surface rises and falls making contact with successive electrodes and completing successive electrical circuits, the water-surface time-history is recorded. The character of this type gage makes it self-calibrating.

CONTINUOUS WIRE GAGE   A vertical wire or pair of wires pierce the water surface. As the surface rises and falls the variation in resistance, capacitance or inductance in the wire(s) is measured to indicate the water-surface time-history. This type of gage requires calibration and like the step-resistance gage, requires a sturdy mounting structure such as a pier or offshore drilling tower.

ACCELEROMETER BUOY   A buoy floats on the water surface and records its vertical component of acceleration. A double integration of the vertical-acceleration time-history yields the water-surface time-history. The buoy is anchored in the area of interest and data are recorded internally or telemetered back to shore.

Wave Hindcasts

Using past synoptic weather charts and the wave prediction procedures discussed in Section 5.4 it is possible to hindcast wave characteristics (H_s, T_s, direction) at selected locations over a sufficient time period to develop a wave climate estimate for that site. As an example, National Marine Consultants (1960a) used 6 hourly U.S. Weather Bureau synoptic weather charts for the Pacific Ocean to develop wave statistics for seven deep water stations along the California coast for a 3-year period, 1956-58. Average monthly and annual frequency distributions of significant height and period for the major compass directions were reported. National Marine Consultants (1960b) also developed the wave statistics for the 10 most severe storms affecting three northern California sites during 1951-70. The significant height and period, range of important periods, and wave direction were given at each site for 6-hr intervals for the duration of each storm.

Visual Observations

There are many visual wave observation programs that yield wave data at sea and in coastal regions. The U.S. Naval Weather Service Command publishes a Summary of Synoptic Meteorological Observations (SSMO) for several oceanic regions around the world. The data are from shipboard observations routinely made by ships in passage. Along with basic meteorological data such as wind speed and direction, the significant wave height and period are reported.

The U.S. Army Coastal Engineering Research Center has a program of Longshore Environmental Observations (LEO), which collects surfzone data (usually daily) at many sites along the ocean and the Great Lakes' coasts. Data include observed wave breaker type, height, period, and direction, as well as longshore current and wind, speed and direction. These data are tabulated, analyzed and made available to anyone who can make use of the data. Some caution should be exercised in the use of SSMO, LEO, and other data from observers as a natural tendency to avoid stormy weather may bias some of the data.

5.4. WAVE FORECASTING

At present, prediction of the spectrum of waves that will be generated by a given wind speed, direction, fetch, and duration is as much an art as it is a science. Further advances, particularly in the understanding of the wave generation mechanism and in the quantity and quality of meteorological and wave data (especially during major storms) are needed to improve wave prediction techniques. The coastal engineer will usually leave wave forecasting to specialists but should have a general understanding of the procedures commonly used.

Prediction of wave characteristics at a particular coastal site requires (1) prediction of the wind waves generated by a storm, (2) evaluation of the changes that occur as swell propagate from the edge of the storm to a deep water point offshore of the site, and (3) evaluation of transitional and shallow water effects (e.g. refraction, diffraction, shoaling) as waves reach the shore. The third item was covered in Chapters 2 and 3 while the first two are covered briefly in this section.

There are two basic deep water wave forecasting methods in common use: the SMB method devised by Sverdrup and Munk (1947) and improved by Bretschneider (1952a, 1971), who analyzed additional data; and the PNJ method from Pierson, Neumann, and James (1955). Both have a theoretical basis but rely heavily on empirical data. The SMB method yields the significant height and period from which the height distribution and spectrum can be evaluated using Eqs. 5.4 and 5.12. The PNJ method uses the Neumann spectrum to directly predict the resulting spectrum of waves. Only a brief discussion of the PNJ method will be presented, while the SMB method, which is more commonly used in engineering practice will be covered in somewhat more detail. Both methods are more thoroughly covered by Kinsman (1965) and Bretschneider in Ippen (1966). Decay of swell, prediction of wind waves generated in shallow water, and hurricane wave prediction will also be briefly covered.

Wave prediction methods require a knowledge of the wind speed, direction, fetch, and duration. For an enclosed water body with a limited fetch such as a coastal bay, wind records from nearby locations may be sufficient to select a design wind speed, direction, and duration. For hurricane waves the wind field of the SPH or PMH can be used. Otherwise, recourse must be made to analysis of synoptic weather charts.

From the isobars (lines of constant pressure) on a weather chart it is possible to calculate the upper atmosphere (geostrophic) wind speed from a static balance between the pressure gradient and Coriolis acceleration. This yields

$$U_g = \frac{(\Delta p / \Delta x)}{2\omega \rho_a \sin \phi} \tag{5.18}$$

where U_g is the geostrophic wind speed generated by a horizontal atmospheric pressure gradient $\Delta p / \Delta x$. The wind would blow approximately parallel to the isobars. Friction causes the surface wind velocity to be less than the geostrophic velocity and to deflect $20°–40°$ from the isobars in the direction of the pressure gradient. The ratio of surface to geostrophic wind speed depends on the air-water temperature difference and is often taken as 0.6 for routine work. From the wind field calculations, the fetches, wave directions, and decay distances for storms that will affect a particular site can be determined. There may be surface wind measurements taken by ships operating in the storm area of interest. These should be used to improve surface wind speed and direction estimates. Weather charts are prepared at 6-hr intervals. Wind durations can be estimated from the observed 6-hourly changes in the wind fields.

SMB Method

In response to a need to forecast sea and swell from weather data during World War II, Sverdrup and Munk (1947) developed a theory for wind wave generation (see Section 5.1) and decay of swell. Both aspects of the theory were verified by the very meager data then available. The resulting wind wave forecasting technique can be demonstrated by the use of dimensional analysis.

In deep water

$$H, C \text{ (or } T) = f(U, F, t_d, g)$$

and dimensional analysis yields

$$\frac{C}{U}\left(= \frac{gT}{2\pi U}\right) = f'\left(\frac{gF}{U^2}, \frac{gt_d}{U}\right)$$

$$\frac{gH}{U^2} = f''\left(\frac{gF}{U^2}, \frac{gt_d}{U}\right)$$

These relationships are plotted in Fig. 5.5 from the original data used by Sverdrup and Munk plus data added by Bretschneider (1952). The data Bretschneider used were from wind-wave flume experiments, measurements made at small lakes, and visual observations and measurements of ocean waves. H and T are the significant wave height and period respectively.

From F and U one can determine the fetch limited significant height and period and from t_d and U one can determine the duration limited significant height and period. The lower set of values would define the resulting wave conditions. For example, consider a 30 m/sec wind blowing over a 200-mile fetch for 8 hr.

$$\frac{gF}{U^2} = \frac{9.81(200)(1854)}{(30)^2} = 4.04 \times 10^3$$

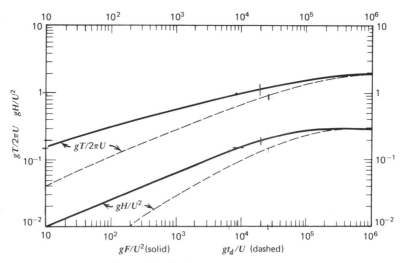

Figure 5.5   SMB wave prediction curves.

From Fig. 5.5 (solid lines)

$$\frac{gH}{U^2} = 0.10 \qquad \frac{gT}{2\pi U} = 0.78$$

For

$$\frac{gt_d}{U} = \frac{9.81(8)(3600)}{30} = 9.42 \times 10^3$$

Figure 5.5 (dashed lines)

$$\frac{gH}{U^2} = 0.094 \qquad \frac{gT}{2\pi U} = 0.64$$

Thus the second or duration limited condition controls and

$$H = H_s = \frac{(30)^2(0.094)}{9.81} = 3.6 \text{ m } (24.2 \text{ ft})$$

$$T = T_s = \frac{2\pi(30)(0.64)}{9.81} = 12.3 \text{ sec}$$

Bretschneider (1971) has plotted the data used for Fig. 5.5 in a dimensional form that is convenient for application.

In Fig. 5.5 the two sets of curves are asymptotic to

$$\frac{gH}{U^2} = 0.282 \qquad \frac{gT}{2\pi U} = 1.95 \tag{5.19}$$

at high values of gF/U^2 and gt_d/U. Equations 5.19 thus define the SMB FDS.

If waves leaving a fetch travel as swell for a decay distance D measured along a great circle path, the deep water travel time t_t is

$$t_t = \frac{D}{C_g} = \frac{2D}{C_0} = \frac{4\pi D}{gT_s} \tag{5.20}$$

For large decay distances, the travel times of various components of the wave spectrum are such that the long period waves will arrive before other waves in the spectrum. Using individual wave periods in place of T_s in Eq. 5.20 would allow one to estimate arrival times of various components of the spectrum.

Using empirical data from Sverdrup and Munk (1947) and Bretschneider (1952a), Bretschneider (1952b) presented the set of curves in Fig. 5.6 for determining the increased significant period T_D and decreased significant height H_D at the end of the decay distance, as a function of the significant period and height (T_F, H_F) at the end of the fetch. F_{min} is the fetch F if conditions are fetch limited. If conditions are duration limited, F_{min} is the fetch required for the waves to grow to their limit (see Fig. 5.2). Some of the data used to develop Fig. 5.6 were measured in shallow water and calculations for shoaling had to be made to transform the data back to deep water conditions. Bretschneider (1965) suggested that data scatter was such that deviations of ± 50 percent can easily occur from values predicted by Fig. 5.6.

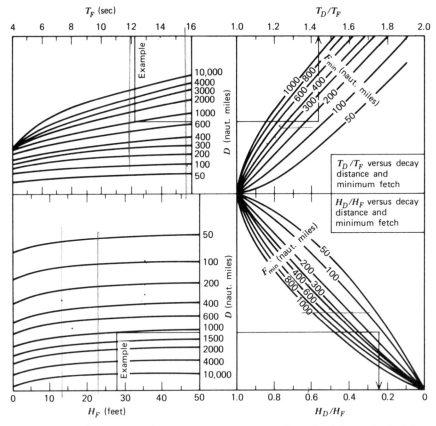

Figure 5.6 H_D/H_F and T_D/T_F versus decay distance D and minimum fetch (after Bretschneider, 1952b).

In the above example at $gH/U^2=0.094$, $gF/U^2=2.5\times10^3$, so

$$F_{min} = \frac{2.5(10^3)(30)^2}{9.81(1854)} = 123 \text{ miles}$$

Thus for a decay distance of, say, 1000 miles Fig. 5.6 yields

$$\frac{H_D}{H_F} = 0.24 \qquad \frac{T_D}{T_F} = 1.43$$

with $H_F=8.6$ m and $T_F=12.3$ sec. Thus the significant height and period at the end of the decay distance are 2.1 m and 17.6 sec. Using the average significant period

$$t_t = \frac{4\pi(1000)(1609)}{(9.81)(14.8)} = 139,200 \text{ sec } (38.7 \text{ hr})$$

PNJ Method

Pierson, Neumann, and James (1955) developed a wave forecasting method based on the Neumann energy spectrum and the relationship between the significant height and total specific energy demonstrated by Longuet-Higgins (1952). Results are presented as co-cumulative spectra (CCS) curves for a range of wind speeds, durations, and fetches. A CCS is the integral from infinity to zero of the energy frequency spectrum. For a given wind speed, the controlling fetch or duration limited CCS is determined and this further yields the significant height and period and the range of important periods.

The PNJ method, which is also semi-empirical, is based on different data than that used by Sverdrup, Munk, and Bretschneider, so for the same wind speed, fetch, and duration slightly different results will be obtained.

Numerical Models

Recently, numerical models (see Barnett, 1966 and Inoue, 1967) have been developed that show excellent potential for improved wave forecasts, particularly where there are significant spatial and temporal wind field variations. Energy transfers into and out of the wave spectrum are computed at a system of grid points for small time increments. The result is the wave spectrum at each grid point and time interval for the duration of the wave generating wind field.

Hurricane Waves

The SMB and PNJ methods could be used to predict the characteristics of waves generated by a hurricane. However, the wind field irregularity in a hurricane is such as to cause the results to be of questionable value, particularly for a fast-moving hurricane. Wilson (1955) developed a graphical technique for predicting waves generated by an irregular moving wind field that is well suited to hurricane wave prediction. The technique is based on the SMB method but can also be adapted to the PNJ method. A basic requirement is a plot, on a position versus time graph, of contours of constant wind velocity component in the direction of interest. This is superimposed on a set of forecasting curves and graphical integration proceeds, yielding significant heights and periods as a function of position and time.

A simple approach for predicting the maximum significant deep water hurricane waves was given by Bretschneider (1957). He presents the empirical formulas

$$H_s = 16.5 e^{0.01 R \Delta p} \left(1 + \frac{0.208 V_F}{\sqrt{U_R}} \right)$$

$$T_s = 8.6 e^{0.005 R \Delta p} \left(1 + \frac{0.104 V_F}{\sqrt{U_R}} \right) \tag{5.21}$$

for the significant wave height and period generated at the point of maximum winds for hurricanes of slow to moderate forward speed. In Eq. 5.21 V_F is the forward speed of the hurricane in knots, U_R is the wind speed in knots at the point of maximum winds, R is the radius to maximum winds in nautical miles, and Δp is the pressure difference between the eye and periphery of the hurricane in inches of mercury. $R \Delta p$ is called the energy index of a hurricane and is an indicator of hurricane magnitude. Bretschneider (see Ippen, 1966) gives a plot of the H_s distribution in a typical hurricane as a function of H_s at R. The plot also shows the directions of wave propagation throughout the hurricane.

Shallow Water

As storm waves propagate over transitional and shallow water depths wave growth is opposed by bottom friction and wave heights are limited by breaking (see Section 2.8). A numerical method for considering bottom friction, refraction, and breaking effects on wave generation in shallow water has been given by Bretschneider (1957). A spatial step calculation is performed with wave growth determined from the SMB method balanced by wave decay due to friction using an approach developed by Bretschneider and Reid (1954). This method is especially useful for predicting hurricane waves in regions having a borad shallow shelf such as the United States Gulf coast. The starting conditions for calculations are the deep water wave characteristics as determined by Section 5.4-Hurricane Waves.

Using the procedures he presented in 1957, Bretschneider developed curves for wave generation in shallow water of various depths. The curves and an example application are given in U.S. Army Coastal Engineering Research Center (1973).

5.5. SUMMARY

Most coastal engineering planning and design requires a thorough knowledge of the wind wave climate that can be expected at a potential site. This chapter addresses part of that requirement by discussing techniques for developing wind wave hindcasts as well as for measuring and analyzing wind wave spectra. This, coupled with the material presented in Chapters 2 and 3 allows the development of the wave climate at a nearshore location. Chapters 6 and 7, which cover wave-structure interaction and coastal zone processes, will rely on a knowledge of the nearshore wave characteristics.

5.6. REFERENCES

Barnett, T. P. (1968), "On the Generation, Dissipation and Prediction of Ocean Wind Waves," *Journal, Geophysical Research*, Vol. 73, pp. 513–530.

Blackman, R. B. and J. W. Tukey (1958), *The Measurement of Power Spectra*, Dover, New York, 190 p.

Bretschneider, C. L. (1952a), "The Generation and Decay of Wind Waves in Deep Water," *Transactions, American Geophysical Union*, Vol. 33, pp. 381–389.

Bretschneider, C. L. (1952b), "Revised Wave Forecasting Relationships," *Proceedings, Second Conference on Coastal Engineering*, Council on Wave Research, Berkeley, pp. 1–5.

Bretschneider, C. L. (1957), "Hurricane Design Wave Practices," *Journal, Waterways and Harbors Division*, American Society of Civil Engineers, Vol. 83, pp. 1238-1–1238-3.

Bretschneider, C. L. (1959), "Wave Variability and Wave Spectra for Wind-Generated Gravity Waves," Tech. Memo. 118, U.S. Army Beach Erosion Board, Washington, D. C., 192 p.

Bretschneider, C. L. (1971), Deep water wave forecasting curves in U.S. Army Coastal Engineering Research Center *Shore Protection Manual*, pp. 3–36, 3–37, 1973.

Bretschneider, C. L. (1965), "Generation of Waves by Wind—State of the Art," Report SN-134-6, Antional Engineering Science Co., Washington, D. C., 96 p.

Bretschneider, C. L. and R. O. Reid (1954), "Modification of Wave Height Due to Bottom Friction, Percolation and Refraction," Tech. Memo 45, U.S. Army Beach Erosion Board, Washington, D. C., 36 p.

Collins, J. I. (1967), "Wave Statistics from Hurricane Dora," *Journal, Waterways and Harbors Division*, American Society of Civil Engineers, Vol. 93, pp. 59–77.

Draper, L. (1966), "The Analysis and Presentation of Wave Data—A Plea for Uniformity," *Proceedings, Tenth Conference on Coastal Engineering*, American Society of Civil Engineers, pp. 1–11.

Draper, L. (1972), "Extreme Wave Conditions in British and Adjacent Waters," *Proceedings, Thirteenth Conference on Coastal Engineering*, American Society of Civil Engineers, p. 157–165.

Goodknight, R. C. and T. L. Russell (1963), "Investigation of the Statistics of Wave Heights," *Journal, Waterways and Harbors Division*, American Society of Civil Engineers, Vol. 89, pp. 29–54.

Grace, R. A. (1970), "How to Measure Waves," *Ocean Industry*, February, pp. 65–69.

Harris, D. L. (1974), "Finite Spectrum Analysis of Wave Records," *Proceedings, Ocean Wave Measurement and Analysis Symposium*, New Orleans, pp. 197–124.

Hasselmann, K. and fifteen co-authors (1973), "Measurement of Wind-Wave Growth and Swell Decay during the Joint North Sea Wave Project (JONSWAP), Report from German Hydrographic Institute, Hamburg, 95 p.

IEEE (1967), Series of papers on the Fast Fourier Transform in *IEEE Transactions on Audio and Electroacoustics*, June, Vol. AU-15, pp. 44–117.

Inoue, T. (1967), "On the Growth of the Spectrum of a Wind-generated Sea According to a Modified Miles-Phillips Mechanism and Its Application to Wave Forecasting," Tech. Rept. 67-5, Geophysical Science Lab, New York University.

Ippen, A. T. (1966), *Estuary and Coastline Hydrodynamics*, McGraw-Hill, New York, 744 p.

Jeffreys, H. (1925), "On the Formation of Water Waves by Wind," *Proceedings, Royal Society*, Series A, Vol. 107, pp. 189–206.

Kinsman, B. (1965), *Wind Waves*, Prentice-Hall, Englewood Cliffs, New Jersey, 676 p.

Longuet-Higgins, M. S. (1952), "On the Statistical Distribution of the Heights of Sea Waves," *Journal, Marine Science*, Vol. II, pp. 345–366.

McClenan, C. M. and D. L. Harris (1975), "The Use of Aerial Photography in the Study of Wave Characteristics in the Coastal Zone," Tech. Memo. 48, U.S. Army Coastal Engineering Research Center, 72 p.

Miles, J. W. (1957), "On the Generation of Surface Waves by Shear Flows," *Journal, Fluid Mechanics*, Vol. 3, pp. 185–204.

National Marine Consultants (1960a), "Wave Statistics for Seven Deep Water Stations Along the California Coast," Santa Barbara, California.

National Marine Consultants (1960b), "Wave Statistics for Ten Most Severe Storms Affecting Three Selected Stations Off the Coast of Northern California, During the Period 1951–1960," Santa Barbara, California.

Neumann, G. (1952), "An Ocean Wave Spectra and a New Method of Forecasting Wind-Generated Sea," Tech. Memo. 43, U.S. Army Beach Erosion Board, Washington, D. C., 42 p.

Peacock, H. G. (1974), "CERC Field Wave Gaging Program," *Proceedings, Ocean Wave Measurement and Analysis Symposium*, New Orleans, pp. 170–185.

Phillips, O. M. (1957), "On the Generation of Waves by Turbulent Wind," *Journal, Fluid Mechanics*, Vol. 2, pp. 417–445.

Pierson, W. J. (1954), "An Interpretation of the Observable Properties of Sea Waves in Terms of the Energy Spectrum of the Gaussian Record," *Transactions, American Geophysical Union*, Vol. 35, pp. 747–757.

Pierson, W. J. and L. Moskowitz (1964), "A Proposed Spectral Form for Fully Developed Wind Seas Based on the Similarity Theory of S. A. Kitaigorodskii," *Journal, Geophysical Research*, Vol. 69, pp. 5181–5190.

Pierson, W. J., G. Neumann, and R. W. James (1955), "Practical Methods for Observing and Forecasting Ocean Waves by Means of Wave Spectra and Statistics," U.S. Navy Hydrographic Office Publ. 603, 284 p.

Ribe, R. L. and E. M. Russin (1974), "Ocean Wave Measuring Instrumentation," *Proceedings, Ocean Wave Measurement and Analysis Symposium*, New Orleans, pp. 396–416.

Sverdrup, H. U. and W. H. Munk (1947), "Wind, Sea and Swell: Theory of Relations for Forecasting," U.S. Navy Hydrographic Office Publ. 601, 44 p.

Thompson, E. F. (1974), "Results from the CERC Wave Measurement Program," *Proceedings, Ocean Wave Measurement and Analysis Symposium*, New Orleans, pp. 836–855.

Thompson, E. F. and D. L. Harris (1972), "A Wave Climatology for U.S. Coastal Waters," *Offshore Technology Conference* paper 1693, Houston, Texas.

Tucker, M. J. (1963), "Analysis of Records of Sea Waves," *Proceedings, Institution of Civil Engineers*, Vol. 26, pp. 305–316.

U.S. Army Coastal Engineering Research Center (1973), *Shore Protection Manual*, 3 vols., U. S. Government Printing Office, Washington, D. C.

Wilson, B. (1955), "Graphical Approach to the Forecasting of Waves in Moving Fetches," Tech. Memo. 73, U.S. Army Beach Erosion Board, Washington, D. C., 31 p.

Yang, C. Y., M. A. Tayfun and G. C. Hsiao (1974), "Extreme Wave Statistics for Delaware Coastal Waters," *Proceedings, Ocean Wave Measurement and Analysis Symposium*, New Orleans, pp. 352–361.

5.7. PROBLEMS

1. Plot $p(H_i)$ and $P(H_i)$ versus H_i for a significant wave height of 5 m. Assume the wave height distribution is defined by a Rayleigh distribution.

2. If the mean wave height in a wave record is 3 m, estimate the percentage of the heights exceeding 4 m.

3. A wind having a speed of 30 knots measured at the standard elevation of 10 m blows for a sufficient duration and over a sufficient fetch for development of a FDS. Assuming a Pierson-Moskowitz spectrum, determine the significant height and period and plot the energy-period spectrum.

4. Wave energy spectra are often given in terms of the angular frequency, $\sigma = (2\pi)/T$. If $S_{H^2}(T)\,dT = -S_{H^2}(\sigma)\,d\sigma$, develop the relationship between $S_{H^2}(T)$ and $S_{H^2}(\sigma)$, and the angular frequency relationship for the Pierson-Moskowitz spectrum.

5. Draw a typical deep water energy-period spectrum and then demonstrate the effects of the following on this spectrum.

   a. Measurement of the wave spectrum by a submerged pressure type wave gage.
   b. Diffraction of normally incident waves behind a breakwater.
   c. Refraction and shoaling as waves propagate into shallow water with a spreading of wave orthogonals.

6. A wind speed of 60 knots blows over a 400-mile fetch for 18 hr. Determine the significant height and period at the end of the fetch (at $t = 18$ hr) and at a decay distance of 800 miles.

7. In the previous Problem, estimate the maximum wave height that would be observed at the end of the fetch.

8. A 50 knot wind blows over a 400-mile fetch for a sufficient duration to be fetch limited. Determine the significant height and period at distances of 100, 200, 300, and 400 miles downwind.

9. Plot the Bretschneider spectrum for each of the four locations in the previous problem.

10. Swell from the South Pacific arrive at the California coast at the time of a local storm. Sketch the form of the energy-period spectrum you would anticipate. Explain.

11. Assume the SPH given in Fig. 4.11 has $R = 26$ n.m. and CPI $= 27.55$ in. of mercury. Use the SMB method to determine the significant height and period of waves generated along a fetch through R parallel to the direction of forward motion. Compare your results with those obtained from Eq. 5.21 if the forward speed of the hurricane is 11 knots. Discuss your results.

12. Analysis of a wave record taken in deep water shows the following joint distribution of wave heights (m) and periods (sec) given as the number of waves in each category. Assuming no refraction or diffraction occurs as waves shoal, determine and plot the cumulative probability distribution of wave runup on a $1:20$ grass slope.

						T			
H	1–2	2–3	3–4	4–5	5–6	6–7	7–8	8–9
0.33–0.67	—	2	3	—	—	—	—	—
0.67–1.00	8	33	16	6	7	—	—	—
1.00–1.33	13	29	44	21	33	12	12	—
1.33–1.67	23	54	83	69	58	42	25	4
1.67–2.00	2	50	87	60	38	35	2	2
2.00–2.33	—	15	29	21	21	4	2	—
2.33–2.67	6	4	19	2	—	—	—	—
2.67–3.00	—	10	2	4	—	—	—	—

13. Plot the cumulative probability distributions for wave height and wave period squared for the data from the previous problem. Also plot the Rayleigh distributions for height and period squared and compare.
14. Waves from a storm are recorded by a gage located offshore in deep water. The average period decreases from 9 sec to 6 sec in 6 hr. How far away from the gage were the waves generated?

Six

WAVE-STRUCTURE INTERACTION

A large segment of coastal engineering design requires an analysis of the functional and structural behavior of a variety of coastal zone structures. Of paramount importance is the response of these structures to wave attack. In certain instances wind, current, earthquake, and ice forces can also be important. Adequate functional design often requires evaluation of the wave reflection and transmission characteristics of structures, particularly in the case of floating and rubble mound breakwaters.

Some wave force problems can be solved as a potential flow problem but, in most instances, empiricism guided by theoretical solutions or dimensional analysis is required.

This chapter covers the forces exerted on, and in some cases the stability of, various classes of structures when subjected to wave attack. These classes include cables, piles, pipelines, large submerged structures, rigid vertical barriers, floating structures, and rubble mound structures. When appropriate, wave reflection and transmission characteristics are presented. Forces on coastal structures induced by wind, currents, ice, and earthquakes are also briefly covered.

6.1. HYDRODYNAMIC FORCES IN UNSTEADY FLOW

For steady flow of an incompressible fluid past a submerged solid object the total drag force F_D caused by frictional shear stress plus the normal pressure distribution can be written

$$F_D = \frac{C_D}{2} \rho_f A U^2 \tag{6.1}$$

In Eq. 6.1, ρ_f is the fluid density, U is the undisturbed flow velocity past the object, A is either the surface area or cross-sectional area normal to the flow depending on whether friction or pressure drag predominates respectively, and C_D is a drag coefficient that depends on the object's shape, orientation, surface roughness, and the Reynolds number.

For a circular cylinder the Reynolds number, $R = UD/\nu$ where D is the cylinder diameter and ν is the fluid kinematic viscosity. Experimental evaluation of C_D as a function of R and the relative roughness, ε/D yields the curves in Fig. 6.1. The size of the surface roughness elements is given by ε. Refer to any standard fluid mechanics text for a physical explanation of the $C_D = f(R, \varepsilon/D)$ curves. Hoerner (1965) gives a thorough compilation of drag coefficients and related information for various shapes.

When flow past an object is unsteady, the total force on the object F at any instant can be written

$$F = \frac{C_D}{2} \rho_f A U^2 + k \mathcal{V} \rho_f \frac{dU}{dt} + \int_A P_x \, dA \tag{6.2}$$

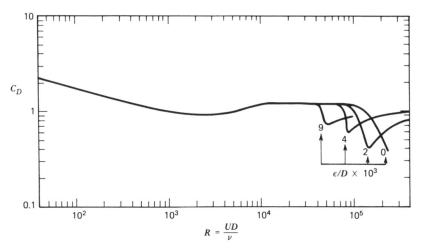

Figure 6.1   Drag coefficient for circular cylinders of varying surface roughness.

The first term on the right is the drag force where the instantaneous value of C_D may not be exactly the same as the steady flow value at the same Reynolds number. The second term accounts for the added mass of fluid set into motion because of the existence of the object. (If, for example, an object at rest in a still fluid is accelerated to some particular velocity, the surrounding fluid that was initially still is also set into motion. A force is required to accelerate this additional mass of fluid.) Ψ is the volume of the object so $\rho_f \Psi$ is the mass of fluid displaced by the object. k is the ratio of the mass of a hypothetical volume of fluid having an acceleration dU/dt to the actual mass of fluid set in motion at its true acceleration. Thus $k\rho_f \Psi$ is the added mass.

Acceleration of the fluid at dU/dt requires a fluid pressure gradient acting in the direction of the acceleration. This pressure gradient also acts on the object. Thus the third term on the right in Eq. 6.2 is the force due to this pressure gradient acting on the object, and is written as the integral of the pressure component in the direction of acceleration over the surface area of the object. It is obvious that this pressure gradient force should be capable of accelerating a mass of the fluid having the same volume as the object at a rate dU/dt. Thus

$$\int_A P_x dA = \rho_f \Psi \frac{dU}{dt}$$

and Eq. 6.2 becomes

$$F = \frac{C_D}{2} \rho_f A U^2 + (1+k)\Psi \rho_f \frac{dU}{dt} \tag{6.3}$$

In potential flow, k has the values given below for various object shapes and orientations to the flow.

Object (and Orientation)	k
Sphere	0.50
Cube—flow normal to a side	0.67
Circular cylinder—flow normal to axis	1.00
Square cylinder—flow normal to axis	1.20

In a real fluid, flow patterns past the object and thus the value of k also depend on surface roughness, the Reynolds number, and the past history of the flow.

Typically, $1 + k$ is called the coefficient of mass or inertia C_M and Eq. 6.3 is written

$$F = \frac{C_D}{2} \rho_f A U^2 + C_M \Psi \rho_f \frac{dU}{dt} \tag{6.4}$$

For potential flow past a circular cylinder, $C_M = 1 + k = 2.0$, but for a real fluid, C_M values less than 2.0 are common (see Sarpkaya and Garrison, 1963, for example). Equation 6.4, when applied to wave forces on submerged objects, is often called the Morison equation after Morison et al. (1950) who first applied it to a study of wave forces on piles.

6.2. PILES, PIPELINES, AND CABLES

Marine piles, pipelines, and cables constitute a class of long cylindrical structures that must be designed to withstand the unsteady flow forces due to wave action. Electrical cables laid along the ocean bottom are somewhat similar to underwater pipelines from a stability point of view, but cables are usually less than 15 cm in diameter while some pipelines such as municipal sewer outfalls can be up to 3 m in diameter. Marine cables are also used to moor ships, buoys, and other floating structures. Piles for piers, offshore drilling structures, dolphins, and so on, are usually vertical or near vertical, but these structures can also have horizontal or inclined cylindrical members for cross bracing. Pile diameters will vary from less than a meter for piers, up to several meters for the legs of some deepwater oil drilling structures.

For a circular cylinder with its axis oriented in a horizontal y or vertical z direction and wave propagation normal to the axis, the force F_s per elemental length ds on the cylinder can be written

$$F_s = \frac{F}{ds} = \frac{C_D}{2} \rho_f D u^2 + C_M \left(\frac{\pi D^2}{4} \right) \rho_f \frac{\partial u}{\partial t} \tag{6.5}$$

The particle velocity u is given by Eq. 2.21, and the local acceleration $\partial u / \partial t$, given by Eq. 2.23, is used in place of the total acceleration du / dt. Use of the local acceleration yields reasonable results for most cases but particularly for waves of low steepness in deeper water.

The water particle acceleration lags the particle velocity by 90° so the drag F_D and inertia F_I components of F_s at a given point will vary as shown in Fig. 6.2. For Fig. 6.2 it was assumed that C_D and C_M are constant and that wave and structure conditions are such to cause the peak inertia and drag forces to be equal. In any given situation the peak total forces occur at

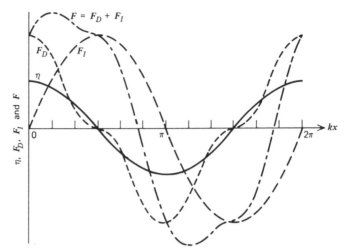

Figure 6.2 Variation of surface elevation and drag, inertia, and total force for equal peak drag and inertia forces.

some instant between arrival of the wave crest or trough and the still water line, the exact position depending on the values of C_D and C_M, the wave height and period, the water depth, and the cylinder diameter. Inserting the relationships for wave particle velocity and acceleration in Eq. 6.4, differentiating F with respect to the phase angle θ ($\theta = kx - \sigma t$), and setting the result equal to zero yields

$$\sin\theta_p = \frac{2C_M \mathscr{V}\sinh kd}{C_D AH\cosh k(d+z)} \tag{6.6}$$

where θ_p is the phase angle for the peak force.

For a given wave condition, F_D is proportional to D and F_I is proportional to D^2, so as the structure size increases, inertia forces tend to dominate and vice versa. Thus for cables, the wave drag force strongly predominates while for large structures such as submerged oil storage tanks the resulting wave force is effectively due solely to inertia effects. For piles, either component can dominate with drag forces being largest at higher H/D ratios and vice versa (see Prob. 6.2).

Often, structures are attacked simultaneously by waves and a current moving at some angle to the direction of wave motion. The drag force on a structure is due to the combined effects of the current and wave particle velocities. The wave characteristics are modified by the current, so the exact nature of the resulting force on a structure is difficult to determine. The usual design procedure is to vectorally add the current and wave particle velocities and use the resulting velocity component in the drag term of the Morison equation.

Piles

The force per unit length from Eq. 6.5 can be integrated from the mud line to the water surface for a vertical pile to obtain the total wave force on the pile as a function of time. Thus

$$F = \int_{-d}^{\eta} \left(C_M \rho_f \frac{\pi D^2}{4} \frac{\partial u}{\partial t} + \frac{C_D}{2} \rho_f D u^2 \right) dz$$

Assuming C_M and C_D are constant, letting $x = 0$ ($kx - \sigma t = \sigma t$), inserting the expressions for u and $\partial u / \partial t$ from Chapter 2, and integrating yields

$$F = \frac{\pi^2 \rho_f D^2 H L C_M}{4 T^2} \frac{\sinh(\eta + d)}{\sinh kd} \sin \sigma t$$

$$+ \frac{\pi \rho_f D H^2 L C_D}{16 T^2} \frac{2k(\eta + d) + \sinh 2k(\eta + d)}{(\sinh kd)^2} (\cos \sigma t)^2 \qquad (6.7)$$

The moment M_0 on the pile around the mudline can be found by integrating the product of the force per unit length and the distance up from the mudline $(d + z)$ over the total submerged length of the pile. Again, holding C_D and C_M constant,

$$M_0 = \int_{-d}^{\eta} \left(C_M \rho_f \frac{\pi D^2}{4} \frac{\partial u}{\partial t} + \frac{C_D}{2} \rho_f D u^2 \right) (d + z) dz$$

or

$$M_0 = - \frac{\pi \rho_f D^2 H L^2 C_M}{8 T^2} \frac{\left(k(\eta + d) \sinh k(\eta + d) - \cosh k(\eta + d) + 1 \right)}{\sinh kd} \sin \sigma t$$

$$+ \frac{\rho_f D H^2 L^2 C_D}{64 T^2} \frac{\left(2k^2(\eta + d)^2 + 2k(\eta + d) \sinh 2k(\eta + d) - \cosh 2k(\eta + d) + 1 \right)}{(\sinh kd)^2}$$

$$\times (\cos \sigma t)^2 \qquad (6.8)$$

Dean and Harleman in Ippen (1966) give the equations for wave force and moment for the deep and shallow water limits and Cross and Wiegel (1965) present computer solutions of Eqs. 6.7 and 6.8 for a range of wave conditions, water depths, and pile diameters.

Important for the calculation of wave forces and moments on piles is selection of values for C_D and C_M. One could use the theoretical value of 2.0 for C_M and, after calculating a Reynolds number from the maximum or rms wave particle velocity, determine C_D from Fig. 6.1. However, since $C_M = 2.0$ is based on potential flow and Fig. 6.1 is for steady flow, a number of experiments have been conducted to further evaluate C_D and C_M for use in wave force analysis. These fall into three categories: a) laboratory measurements of the force on cylinders subjected to constantly accelerating flow, b) measurement of forces exerted on cylinders by monochromatic waves in laboratory flumes, and c) measurement of wave forces on instrumented piles under natural conditions at sea.

Constant fluid acceleration differs from wave particle motion where the acceleration continually changes in a cyclic pattern. Also, in a wave field eddies generated by flow past a pile are then swept back past the pile to affect the subsequent flow pattern around the pile and resulting values of C_D and C_M. For constant acceleration, Eq. 6.5 can be written

$$\frac{F_s}{\rho_f D u^2} = \frac{C_D}{2} + \frac{\pi}{2} C_M \left[\frac{D \left(\partial u \right)/\left(\partial t \right)}{u^2} \right] = C \tag{6.9}$$

where the resistance coefficient C expresses the combined effects of C_D and C_M and is also dependent on the term in brackets, which has been called the Iverson number. Studies of cylinders in constantly accelerating flow by Kiem (1956), Laird et al. (1959) and Sarpkaya and Garrison (1963) have reported their results in terms of C as well as C_D and C_M. Results of some pile wave force studies (Crooke, 1955 and Keulegan and Carpenter, 1958) have also been presented in terms of C.

Figure 6.2 demonstrates that F_I is zero at a wave crest and trough so $F_s = F_D$; and at the midpoints between the crest and trough F_D is zero so $F_s = F_I$. Thus, using the measured height and period in Eq. 2.21 to compute u, and the measured force at the crest or trough, one can evaluate C_D from the drag term of the Morison equation. Likewise, from the measured forces at the midpoint and a computed value of $\partial u / \partial t$ from Eq. 2.23 one can evaluate C_M from the inertia term of the Morison equation. This is the most common procedure for determining C_D and C_M for most laboratory and some field measurements of the time history of wave force on a pile and the water surface time history at the pile.

Laboratory and field evaluations of C_D and C_M in the above manner often exhibit a great deal of scatter of the resulting values. Part of the scatter is due to inaccuracies in measuring $F(t)$ and $\eta(t)$. $\eta(t)$ should be evaluated at the pile without any disturbing effects of the pile. Another contributing factor is errors in predicting u and $\partial u / \partial t$ from $\eta(t)$, particularly since u is squared in the drag term. It is recommended that the same wave theory used in evaluating C_D and C_M from experimental data be used in subsequent wave force predictions for design.

From previous discussion it should be evident that C_D and C_M are not constant through a wave cycle. The values determined by the above procedure are instantaneous values, which are not necessarily the average or maximum values that occur. The peak total force on a pile does not occur at the phase angle where C_D and C_M were determined unless either F_D or F_I strongly predominate.

Although laboratory studies of wave forces on piles allow for greater control of the wave train and easier measurement of the wave surface elevation and wave force time histories, laboratory scale effects raise some questions as to the complete applicability of results to prototype conditions. Reynolds numbers for most laboratory studies fall below 5×10^4, while Reynolds numbers for most prototype conditions of interest are between 10^5 and 10^7.

Numerous laboratory studies of wave forces on circular cylindrical piles have been conducted (see Morison et al., 1950; Harleman and Shapiro, 1955; Keulegan and Carpenter, 1956; Paape and Breusers, 1966; and Jen, 1968).

Keulegan and Carpenter (1956) found that C_D and C_M correlated better with the dimensionless ratio

$$\frac{u_m T}{D}$$

where u_m is the maximum value of the horizontal particle velocity component, than with the Reynolds number. Using the linear wave theory it can be shown that

$$\frac{u_m T}{D} = \frac{\pi (2\zeta)}{D}$$

where 2ζ is the total particle horizontal displacement.

Considering the difficulties encountered in using the Morison equation, Paape and Breusers (1966) recommended that the experimental problem of wave forces on piles be approached by dimensional analysis, yielding

$$\frac{F}{\rho_f g H D^2} = f\left(\frac{d}{gT^2}, \frac{H}{gT^2}, \frac{H}{D}\right)$$

One advantage to this approach is that each of the terms can be measured directly so there is no need to resort to particle velocity and acceleration calculations, as is necessary with the Morison equation.

Several field measurements of wave forces have been made with instrumented piles installed at the end of a pier or on an offshore oil drilling structure. These include studies reported by Wiegel et al. (1957), Reid (1958), Wilson (1965), Evans (1969), and Aagaard and Dean (1969). In addition to the problems experienced in laboratory studies (except for Reynolds number limitations), the irregular characteristics of waves encountered in the field further complicates the determination of C_D and C_M. Also, significant ocean currents may be exerting an additional force on the pile. The field study by Wiegel et al. (1957) produced individual C_D values ranging from 0.1–6.0 and C_M values from 0.4–6.7. Scatter of this magnitude is not uncommon in most field and laboratory studies.

One approach that is useful in evaluating field data to determine C_D and C_M is to employ a least-squares fitting procedure with the Morison equation as the regression relation. In other words, the quantity

$$\frac{1}{T_*} \int_0^{T_*} \left(F - \frac{C_D}{2}\rho_f A u^2 - C_M \rho_f \mathcal{V} \, \partial u/\partial t\right) dt$$

is minimized, where T_* is the length of the wave profile and wave force records. This yields "best fit" values of C_D and C_M for the entire record. Special filtering methods (e.g. see Reid, 1958) to determine u and $\partial u/\partial t$ from the $\eta(t)$ record for the irregular wave field have been valuable in improving the determination of C_D and C_M.

Agerschou and Edens (1965) reanalyzed the data of Wiegel et al. (1957) and some unpublished data from Bretschneider and presented probability distributions of C_D and C_M for various ranges of Reynolds number and wave steepness. Their recommended results along with a tabulation of suggested and average values of C_D and C_M from other laboratory and field studies are presented in Table 6.1.

The particular values of C_D and C_M selected for design use will depend on several factors including the confidence in the selected design wave, the wave theory used, the ratio of F_I to F_D, and the desired factor of safety.

TABLE 6.1.   SUMMARY OF SUGGESTED OR AVERAGE C_D AND C_M VALUES FOR CIRCULAR CYLINDERS

Source	Reynolds Number	C_D	C_M	Remarks
Morison et al. (1950)	$< 2 \times 10^4$	1.60	1.10	Model—waves
Keim (1956)	—	1.0	0.93	Model—constant acceleration
Keulegan and Carpenter (1958)	$< 10^4$	1.34 1.52	1.46 1.51	Model—oscillatory flow
Jen (1968)	$< 2.5 \times 10^4$	—	2.04	Model—waves with F_I predominant
Reid (1958)	5×10^4 to 1.2×10^6	0.53	1.47	Ocean Waves
Wilson (1965)	—	1.00	1.45	Ocean Waves
Agerschou and Edens (1965)	3×10^4 to 9×10^5	1.00 to 1.40	2.00	Ocean Waves
Aagaard and Dean (1969)	2×10^4 to 6×10^7	0.60 to 1.00	1.50	Ocean Waves

Unless these factors dictate otherwise, recommended values for design are $C_D = 1.0$ and $C_M = 1.5$ (without allowance for a factor of safety).

Pile surface roughness can affect the flow field around a pile and the resulting drag (see Fig. 6.1) and inertia coefficients, particularly for the range of Reynolds numbers common to field conditions. Unfortunately, no data are available to indicate the magnitude of surface roughness effects for ocean wave conditions. Most of the prototype drag and mass coefficients reported in Table 6.1 must have included some surface roughness effects but actual or relative surface roughness was not documented in any of the studies. Blumberg and Rigg (1961) evaluated C_D for a 3-ft diameter cylinder with various surface roughnesses, towed through water at a constant velocity. Reynolds numbers for the experiments varied between 1×10^6 and 6×10^6 and C_D increased from 0.58 for a smooth cylinder to 1.02 for the roughest cylinder having oyster shells and concrete fragments in bitumastic on the surface.

Most pile supported ocean structures have several modes of oscillation with typical resonant periods of 1–2 sec or less. Thus their resonant periods are far below the range of higher energy periods in a typical design wave spectrum, and resonant interaction between the structure and ocean waves is not a problem. If resonance is anticipated, as is possible for some of the larger structures in deep water, increased wave loadings will occur and a dynamic analysis of the structure is required.

In addition to resonant response at the incident wave periods, resonance can develop in response to the periodic shedding of vortices generated by wave induced flow past a pile. The period of vortex shedding T_e is given by the Strouhal number

$$S = \frac{D}{T_e u}$$

where u is the steady flow velocity past a cylinder. S for the common range of prototype pile Reynolds numbers varies from 0.2–0.4. For example, given typical values of $D = 1.0$ m, $u = 1.5$ m/s, and $S = 0.3$, $T_e = 2.2$ sec. Thus with a wave period much greater than 2.2 sec, flow past the pile could last long enough for a few cycles of vortex shedding.

Vortex shedding causes oscillatory forces that act normal to the direction of flow. Wiegel and Delmonte (1972) report on wave tank measurements of the oscillating normal or lift forces exerted on a circular cylindrical pile by regular waves. The lift forces often were equal in magnitude to the longitudinal forces exerted in the direction of wave motion and showed a good correlation with the Keulegan-Carpenter number, $(U_m T)/D$.

Pipelines

If possible, particularly in shallow water, pipelines should be buried in the ocean bottom to eliminate damage from waves and currents as well as from dragging ship anchors and trawls and dredging operations. However, during storms bottom scour may be sufficient to expose a pipeline and in many locations bottom conditions may prohibit pipeline burial or even effective anchoring. Then a pipeline must depend on its submerged weight and bottom friction to resist sliding or rolling due to wave and current forces.

When a pipeline is on or near the ocean bottom, the bottom affects the pattern of flow past the pipe and thus the resulting hydromechanic forces on the pipe. The bottom affects flow both by its proximity to the pipe and by its surface characteristics, which cause a boundary layer to form in the flow field around the pipe. Figure 6.3 shows an axial view of a circular pipe resting on a horizontal bottom along with the typical forces acting on a pipe, including those generated by the flow normal to the pipe axis. The effective weight of the pipe is its weight in air W minus the buoyant force F_B. At the pipe-bottom interface there is a normal reaction N and a frictional resistance

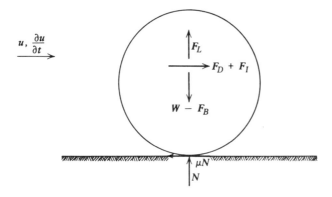

Figure 6.3 Circular pipe—axial view.

μN where μ is the coefficient of sliding friction. The combined wave and current motion cause a horizontal force composed of drag and inertia components that can be evaluated from the Morison equation if the appropriate values of C_D and C_M for this type of flow field can be determined.

Flow past the pipe also generates a lift force having a component similar to the horizontal drag component and due to the distribution of pressure and shear forces on the pipe. There is also an inertial component due to the vertical asymmetry in the acceleration of flow past the pipe. In addition, there may be a fluctuating lift component due to vortex shedding similar to that occurring with flow past a pile. Usually, the lift force is simply written in terms of a lift coefficient C_L as

$$F_L = \frac{C_L}{2} \rho_f A u^2$$

This greatly simplifies a more complex situation. The lift force decreases the normal reaction at the bottom and thus the resulting bottom friction resistance. This, in turn, diminishes the stability of the pipe to sliding, particularly since the unsteady lift force given by Eq. 6.9 is in phase with the horizontal drag force.

If the pipe is raised from the bottom, the vertical asymmetry of flow decreases as does the lift force. When the clearance between the pipe and ocean bottom is about 0.5 pipe diameters, the lift force usually becomes very small.

Several laboratory investigations of the hydrodynamic forces exerted on pipelines laying on the sea floor or suspended just above the floor, have been conducted. The results are usually presented in terms of C_D, C_M, and C_L for use in Eqs. 6.5 and 6.9. Most of the investigations were only for steady flow, particularly those involving higher ranges of Reynolds number, but a few studies in laboratory wave tanks have been conducted. The author is not aware of any prototype scale studies of wave forces on submerged pipelines.

Based on potential flow considerations, published coefficient data for piles, and physical reasoning, Wilson and Reid (1963) recommend use of $C_M = 2.5$, $C_D \geqslant 1.0$, and $C_L \geqslant 1.0$ for a circular pipe resting on the sea floor.

Brown (1967) and Beattie et al. (1971) conducted tests in steady open channel flow to measure drag and lift forces on pipelines by integrating pressures measured around the periphery of a pipe. The pipe used by Brown had a smooth surface, while Beattie et al. used both smooth pipes and pipes roughened by covering their surface with 50-grit sandpaper. Helfinstine and Shupe (1972) evaluated lift and drag forces from strain gage measurements for a pipe in a wind tunnel with steady flow. Below is a summary of results obtained.

Source	Reynolds Numbers	C_D	C_L	Remarks
Brown (1967)	0.6×10^5 to 3×10^5	0.95–0.55	1.3–0.8	Smooth
Beattie et al. (1971)	0.6×10^6 to 2×10^6	0.45–0.65 0.60–0.70	0.65 0.5–0.45	Smooth Rough
Helfinstine and Shupe (1972)	5×10^4 to 1.3×10^5	0.8–0.9 0.8–1.0	1.4 1.2–1.4	Smooth Rough

The results of Beattie et al. and Helfinstine and Shupe suggest that increased pipe surface roughness increases C_D and decreases C_L. Thus increased roughness would increase the drag force on the pipe, but it would also increase the bottom frictional resistance (both μ and N), so the pipeline stability might conceivably remain about the same. Also, the higher the Reynolds number, the lower the values of C_D and C_L reported. The Reynolds number range studied by Beattie et al. is closest to that common to prototype conditions.

Grace (1971) analyzed some Swedish and British data from wave tank studies of forces on pipelines. He also conducted similar wave tank studies. Based on this work he recommends as conservative design values that

$C_D = 2.0$, $C_L = 3.0$ and $C_M = 2.5$. Also, C_D should remain constant for any spacing between the pipe and ocean bottom, C_L should diminish to about 0.9 as the spacing approaches half the pipe diameter, and C_M should "drop off" as the spacing increases.

Brater and Wallace (1972) conducted wave tank tests of wave forces on pipelines in which the wave-pipe characteristics were such that the inertia force predominated. Reynolds numbers ranged between 2×10^3 and 2×10^4. For the condition of the pipe laying on the bottom they found $C_M = 2.59 +$ $4.83(d/L)$, which yields $C_M = 2.8$–3.2 for common transitional to shallow water wave depths. They also presented some results for pipelines in open trenches.

Cables

Flexible rope and steel cables are used for ocean towing and the mooring of such floating structures as ships, buoys, and oil drilling rigs. Hydrodynamic forces from waves and currents are similar to those on cylindrical piles with the additional factor that the cable usually will undergo a series of complex unsteady motions, so relative fluid-structure velocities and accelerations must be considered. Also, the mooring cable tethers a moving body so analysis of the coupled motion of the mooring line and floating body is often required. For a single mooring line on a large vessel the line's influence on the vessel is small and the vessel can be considered to be free of restraint except for resistance to steady drift. The free motions of the vessel then define an end boundary condition on the mooring cable.

Analysis of cable behavior is usually accomplished by some finite element numerical method in which short segments of the cable are analyzed as individual connected members. Besides the hydrodynamic forces on each segment, the other forces are the submerged weight of the segment and the tension on each end. It is usually assumed that the cable is totally flexible and thus cannot develop internal bending moments. Thus there is no shear force at the end of each segment. Depending on the weight and elastic characteristics of the cable it may effectively be extensible or inextensible. The analysis must accordingly use or omit the axial stress-strain relationship. Wilson (1967) presents a discussion of the relevant properties of mooring cables.

The simpler approach to the investigation of cable behavior is the static analysis, in which the resulting cable configuration and tensions are determined from a static balance of the forces acting on each segment. Cable inertia, unsteady wave forces, and the hydrodynamic effects of cable motion are, of course, not included. For examples, see Wilson and Garbaccio (1967),

Dominquez and Filmer (1971), and Chang and Pilkey (1971). A true dynamic analysis of cables, including unsteady hydrodynamic forces and cable inertia, is a difficult problem to solve. Techniques for dynamic analysis of cables have also been presented by Wilson and Garbaccio (1967) and Chang and Pilkey (1971).

6.3. LARGE SUBMERGED STRUCTURES

The recent increase in offshore oil production has created a need for large offshore structures to store crude oil awaiting shipment to refineries. For example, a 500,000 bbl oil storage tank has been placed 97 km offshore in the Persian Gulf where the water depth is 50 m (see Chamberlain, 1970). The tank has the shape of a circular television picture tube with the screen facing down and the neck piercing the water surface. The oil storage section has an 82 m diameter and a crown elevation 24 m above the sea floor. A 100-year wave 12.0 m high having a 10.5-sec period (163 m wave length) was used to calculate design loads for the tank. Design of such structures requires an analysis of the horizontal and lift forces as well as the pressure distribution over the structure surface caused by wave and current motion. When filled with oil the structure is buoyant and must be anchored to the sea floor with piles.

The water-particle horizontal orbit dimension is proportional to the wave height. When this dimension is less than about half the size of a structure there is little boundary layer development or flow separation with trailing vortices. Thus if the ratio H/D is small, as is the case for most large submerged structures, viscous effects become negligible and the total wave force on the structure is due essentially to inertial effects. (This trend is indicated by the Morison equation as discussed in Sect. 6.2.) Negligible viscous effects mean that quite accurate, strictly analytical solutions become possible and viscous scale effects are not important in hydraulic model studies of wave-structure interaction.

When the ratio of structure size to wave length is small the wave is relatively undisturbed as it propagates past a structure. This assumption is inherent in the use of the Morison equation. However, when the structure size is of the same order of magnitude or larger than the wave length (as is the case for large oil storage structures) the water particle motion and wave free surface form are greatly disturbed by the structure. The degree of disturbance of the wave motion also depends on the ratio of vertical structure dimension to water depth.

Hydraulic model studies have been conducted to determine maximum horizontal and vertical components of wave force as well as surface pressure distributions on a variety of common structure shapes. Maximum horizontal and vertical force data are usually presented in terms of

$$\frac{2F_{max}}{\rho_f g a^2 H} = f\left(\frac{2\pi a}{L}, \frac{d}{a} \right)$$

where the maximum force is represented by F_{max}, and a is some significant structure dimension. Garrison and Rao (1971) have presented force data for submerged hemispheres, Chakrabarti and Tam (1973) for a vertical circular cylinder, and Herbich and Shank (1970) for a variety of rectangular shapes.

Analytical methods for determining the pressure distribution and resulting component forces due to wave motion acting on a large submerged structure require use of a diffraction theory. This involves solution of the Laplace equation with appropriate boundary conditions to yield a velocity potential composed of two parts—one representing the incident wave and the other the scattering of the incident wave due to interaction with the structure.

MacCamy and Fuchs (1954) applied the diffraction theory to determine the wave forces on a vertical circular cylindrical structure. When the cylinder size is very large compared to the wave length the results reduce to the inertia term in the Morison equation with $C_M = 2.0$ (i.e. $C_D = 0$) as should be expected for potential flow. Garrison and Rao (1971) have applied the diffraction theory to a submerged hemisphere and Garrison and Chow (1972) present a diffraction theory for submerged objects of arbitrary shape that can be solved numerically by digital computer. Chakrabarti (1973) has presented a simplified approach in which the pressure field in the undisturbed wave motion is integrated over the structure surface and an experimental coefficient is added to account for scattering effects. Results are presented for several basic shapes.

6.4. FLOATING STRUCTURES

Types of floating structures of interest to coastal engineers include oil exploration drilling rigs, buoys for marine data collection and marking navigation routes, and breakwaters. With the recent rapid growth in the pleasure boat market, floating breakwaters are gaining in importance for the protection of marinas located on reservoirs and other semi-protected, relatively deep waters. For reasons to be demonstrated, floating breakwaters usually do not perform satisfactorily when subjected to the longer period waves found on the open coast.

Some of the advantages of floating breakwaters are:

1. They are easily adapted to large water level fluctuations as occur on some reservoirs and their cost does not rapidly increase with increased water depths.
2. They offer less obstruction to water circulation, fish migration, and so on.
3. They are mobile and can be relocated, if necessary.
4. They are less dependent on bottom soil conditions.

Primary concerns include floating structure motion in response to wave action, resulting horizontal and vertical forces exerted by the floating structure on mooring lines, and wave energy transmission past floating structures. These concerns will be discussed in subsequent parts of this section.

The kinetic energy at a point in a wave is proportional to the water particle velocity squared, and thus is strongly dependent on vertical position for deep and transitional water waves. For a deep water wave with $d/L = 0.5$ over 70 percent of the kinetic energy is concentrated in the top 20 percent of the water column. For shallow water waves, on the other hand, the water particle velocity and kinetic energy are essentially constant from the water surface to the bottom. Thus floating structures will interact with a much higher percentage of the kinetic energy in a deep water wave than a shallow water wave.

Figure 6.4 shows a floating structure subjected to an incident wave of height H_i. A portion of the incident wave energy is reflected H_r, a portion is transmitted past the structure H_t, a portion is dissipated at the structure, and the remaining energy sets the structure into motion which, in turn, generates

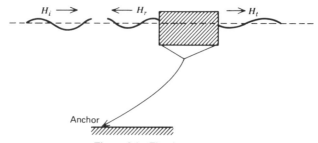

Figure 6.4 Floating structure.

wave motion in the forward and aft directions (included in H_r and H_t). Possible modes of energy dissipation include turbulence in wave breaking on or over the structure, frictional dissipation through surface shear stress and eddy generation, and inelastic deformation of the floating structure. A floating breakwater must reflect and/or dissipate a significant portion of the total incident energy to be effective.

The incident wave power must equal the sum of the power reflected, transmitted, and dissipated. Thus for a unit width of floating structure from Eq. 2.35

$$(nH^2L)_i \geqslant (nH^2L)_r + (nH^2L)_t$$

depending on the amount of energy dissipated. If the water depth is constant

$$H_i^2 \geqslant H_r^2 + H_t^2$$

or

$$1 \geqslant \left(\frac{H_r}{H_i}\right)^2 + \left(\frac{H_t}{H_i}\right)^2 = C_R^2 + C_T^2 \tag{6.10}$$

where C_R and C_T are the reflection and transmission coefficients respectively.

The response of a spring-mass system to harmonic excitation shown in Fig. 4.6 also describes the response of a floating structure to waves. Incident waves with periods at or near the resonant period of the floating structure will cause an amplified response oscillation, the degree of amplification depending on the exact excitation period and the level of hydrodynamic damping of the structure and mooring line. Thus the amplitude of horizontal and vertical motion of the floating structure can be up to several times greater than the incident wave amplitude. If the incident wave period is close to, but less than, the resonant period, there will be a phase lag between the wave and resulting structure motion that can approach 180°. This phase lag will cause a phase lag between the incident and regenerated wave motions thereby increasing the complexity of the reflected and transmitted wave characteristics.

The reflection coefficients for both a fixed and a moored rectangular floating structure, as a function of incident wave length, are shown in Fig. 6.5 (from Kincaid, 1960). The structure had a rectangular cross-section with a width λ and draft $\lambda/2$, and was ballasted to have two different natural periods. Neglecting energy dissipation, a conservative $C_T = \sqrt{1 - C_R^2}$ from Eq. 6.10. For the fixed condition, as the incident wave length increases, the

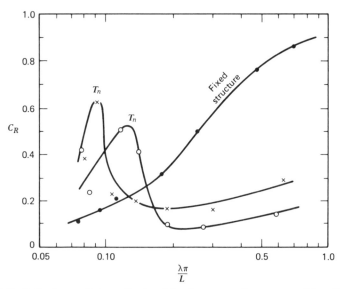

Figure 6.5   Reflection coefficients, fixed and floating rectangular structure (Kincaid, 1960).

reflection diminishes and the transmission increases. When the incident wave length excedes 10 or 15 times the length of the structure, the structure is essentially transparent to the wave (e.g. when $L/\lambda = 15$, $\pi\lambda/L = 0.21$, $C_R = 0.39$, $C_T = 0.92$). A similar trend developed for a moored floating structure but C_R is lower for most relative incident wave periods; the exception being for incident periods near the resonant periods, where structure response markedly increases C_R.

The horizontal force on the moored floating structure increased with increasing wave length to a maximum at the natural period. This was followed by a decrease in force with increasing wave length as the structure became increasingly transparent to the incident waves. Experiments (Harleman and Shapiro, 1961) to measure the horizontal wave force on a fixed sphere and a submerged moored floating sphere showed the maximum force to be greater on the fixed sphere, except in the range $T = (0.5–1.5)\,T_n$. The maximum force on the moored sphere was about three times that on the fixed sphere at the resonant period.

A variety of floating breakwaters have been proposed (see Kowalski, 1974). They exhibit a range of effectivenesses as long as the incident wave length is no greater than a few times the major dimension of the structure. Chen and Wiegel (1974) report on laboratory experiments on an all-purpose, composite floating breakwater that had a vertical barrier to reflect wave energy, a sloping front face to cause waves to break, small openings to dissipate energy by wave generated reversing flow through the walls, and separated air flotation chambers to increase floating stability. When the wave length was equal to the major structure length in the direction of wave propagation $C_T = 0.1-0.2$, but when the wave length was increased to four times the structure length C_T increased to 0.6. A threefold increase in incident wave height for a given wave length caused no detectable increase or decrease in C_T.

Seymour and Isaacs (1974) have proposed a breakwater system composed of a large number of independently tethered floats each having a character-istic dimension about equal to the design incident wave height. The system relies solely on energy dissipation to reduce wave heights. This is accomplished by adjusting the float size, mass, and mooring line length to yield a resonant period in the range of incident wave periods. The amplified float motions that are also out of phase with the water motion (see Fig. 4.6) produce high relative flow velocities. Since the power dissipated by hydrodynamic drag is proportional to the relative velocity cubed, high energy dissipation occurs.

6.5. RUBBLE MOUND STRUCTURES

Rubble mound structures consisting of graded layers of stone and an armor cover layer of stone or specially shaped concrete units are employed in the coastal zone as breakwaters, jetties, groins, and shoreline revetments. One advantage of rubble-mound structures is that failure of the armor cover layer is not sudden, complete, and due to a few large waves but gradual, usually partial in extent, and spread over the duration of the storm. If damage does occur, the structure continues to function and the damage can be repaired after the storm abates, during a period of lower waves. In some cases it may be economical to use smaller size armor units, anticipate a certain degree of damage during a design storm, and provide for subsequent repair of the structure.

Armor units must be of sufficient size to resist wave attack. However, if the entire structure consists of units this size, the structure would allow extremely high wave-energy transmission and finer material in the foundation or embankment being protected could easily be removed. Thus the structure unit sizes are graded, in layers, from the large exterior armor units to small quarry-run sizes and finer at the core and at the interface with the native soil bed.

Other rubble-mound structure design considerations include: prevention of scour at the seaward toe of the structure caused by wave agitation;

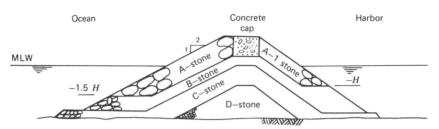

Figure 6.6 Typical rubble-mound breakwater cross-section.

spreading of the structure load so there is no foundation failure owing to excessive loads; and providing sufficient crest elevation and width so wave runup and overtopping do not cause failure of the armor units on the leeward side of the structure or regeneration of excessive wave action in the lee of the structure. The crest width may be dictated by the minimum roadway width needed for construction vehicles that have to traverse the structure.

Figure 6.6 shows a typical breakwater cross-section. Jetties and groins are usually similar, but somewhat less complex. There are many variations in cross-section designs, the variations depending on wave climate, breakwater orientation, water depth, stone availability, foundation conditions, and degree of protection required of the structure. Class A armor stones might each weigh as much as 10 to 25 tons or more, depending on the level of wave attack. The armor layer should extend below the minimum SWL to a depth

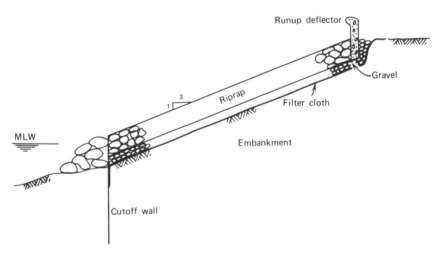

Figure 6.7 Riprap shore revetment.

of 1.5 times the design wave height at the structure. Quarry run D-stone in the core will typically have unit weights of 100 pounds and less and intermediate layer stone sizes will be graded between these limits. Note the extension of the C-stone to assist in toe protection.

A typical riprap shore revetment is shown in Fig. 6.7. Several varieties of interlocking artificial units or a monolithic reinforced concrete slab have also been used. The vertical sheet pile cutoff wall is provided for toe protection from scour and a filter cloth is used to protect the embankment from leaching of fine soil particles. Details of the many variations of rubble mound structures and additional features are given by the U.S. Army Coastal Engineering Research Center (1973).

Armor Unit Stability

The following is a listing and brief discussion of the major factors that control armor unit stability under wave attack.

Incident Wave Spectrum

Since failure of rubble-mound structures is gradual, the significant wave height is most commonly used in design formulas, although more conservative heights such as H_{10} have been used. Some consideration should be given to the expected duration of wave attack when selecting a design wave height. It is also important to determine if the design wave will break on or before the structure or if the water depth is sufficient for the wave to reflect without breaking. If breaking on the structure does occur, armor unit stability is then dependent on the type of breaker (see Ahrens, 1975) which, in turn, depends on the wave height and period and the structure slope.

Armor Unit Size, Weight, Shape, Location, and Method of Placement

Armor unit stability formulas give the weight of a unit required for stability. The resulting size is then dependent on the rock or concrete specific weight. Resistance to hydrodynamic forces is also developed by unit interlocking, which depends on the unit shape and size gradation and the method by which the units are placed during construction. One of the goals of artificial concrete armor unit design is to develop shapes that exhibit a high degree of interlocking while still retaining sufficient porosity when in place. Armor unit stability also depends on location in the breakwater, as exposure to wave attack is usually greater at the head of a breakwater than at some point along the trunk.

Armor Layer Thickness, Porosity, and Slope

Two layers of armor units are usually used to achieve an optimum trade off between initial and reserve stability, prevention of removal of smaller sizes from the under layer, and structure costs. Layer porosities usually vary between 35–55 percent, depending on the armor unit shape and placement method. Low porosities increase the level of wave reflection, an effect that can be very undesirable in certain situations. Low porosities also cause increased wave runup, as well as internal pressure buildup due to the return flow of wave runup. Internal pressure buildup contributes to armor unit instability. Breakwater armor units are all of one or a small range of sizes (usually within ± 25 percent of the average size), but stone riprap revetments often have a much larger size range. The size range of successive layers within a breakwater should increase to decrease breakwater permeability. Typical seaward breakwater slopes vary from 1 on 1.5 to 1 on 3, with revetment slopes being as flat as 1 on 5. A flatter slope increases armor unit stability. It may also increase costs since more material is required even though runup is lower and thus a lower crest elevation may be used. An economic trade off between unit size (layer thickness) and slope length can often be made. Depending on the degree of wave overtopping anticipated, the leeward slope of a breakwater can be steepened to near the angle of repose of the cover layer units (usually 1 on 1.25 as a limit).

Allowable Damage

The degree of damage is usually defined as the percent damage based on the volume of armor units displaced in the zone of wave attack. A certain amount of initial settling of armor units increases the stability of the armor layer. Allowance of up to 30 or 40 percent damage for a design wave will significantly decrease the required armor unit size. However, the damage should not be allowed to develop to the extent that interior layers are exposed to direct wave attack. The allowable damage should depend on initial costs versus maintenance costs, as well as on the allowable risk to areas protected by the structure.

Determination of Armor Unit Stability

The hydrodynamic and unit interaction forces are so complex that a completely analytical development of an armor unit stability equation is not practicable. From a combination of analytical reasoning, dimensional considerations, and experimental data evaluation, the equation (Hudson, 1959)

$$W = \frac{\gamma_r H^3}{K_D \left(S_r - 1\right)^3 \cot \alpha} \tag{6.11}$$

has been developed for W, the weight of an individual stable armor unit. S_r and γ_r are the armor unit specific gravity and specific weight, respectively, and α is the armor unit layer slope. K_D is a dimensionless, experimentally evaluated, coefficient that depends primarily on armor unit shape and method of placement, wave breaking conditions, degree of wave overtopping, and allowable damage.

Most laboratory studies to evaluate K_D have used waves of constant period and height. For irregular waves, it is felt that the significant height is the most appropriate wave height to use for H in Eq. 6.11. There have only been a limited number of evaluations of Eq. 6.11 using irregular waves (Rogan, 1969; Ouellet, 1972); more research with a variety of wave spectra is needed. The only effect of wave period on Eq. 6.11 is in its effects on K_D through the breaking condition. Note that the required unit weight is a function of the wave height cubed so armor unit weights increase rapidly with increased design wave height.

Some values of K_D are listed in Table 6.2 as a function of unit shape, location on the structure, and exposure to breaking or nonbreaking waves.

TABLE 6.2.   SUGGESTED K_D VALUES FOR VARIOUS ARMOR UNIT SHAPES

Armor Unit	Structure trunk		Structure head	
	Breaking Wave	Nonbreaking Wave	Breaking Wave	Nonbreaking Wave
Quarrystone				
Smooth rounded	2.1	2.4	1.7	1.9
Rough angular	3.5	4.0	2.5	2.8
Tetrapod	7.2	8.3	5.5	6.1
Tribar	9.0	10.4	7.8	8.5
Dolos	22.0	25.0	15.0	16.5
Hexapod	8.2	9.5	5.0	7.0
Riprap, graded				
and angular	2.2	2.5	—	—

From: U.S. Army Coastal Engineering Research Center, 1973.

Plate 6.1 Concrete seawall with rubble toe protection, Galveston, Texas. (Courtesy of U.S. Army Coastal Engineering Research Center)

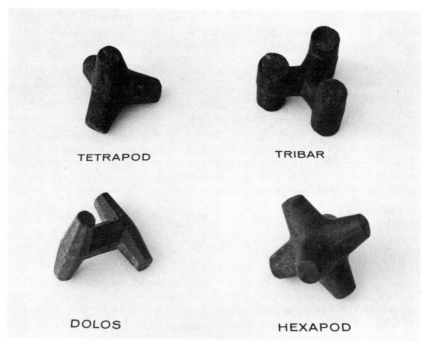

TETRAPOD

TRIBAR

DOLOS

HEXAPOD

Plate 6.2 Typical artificial armor unit shapes. (Author's photo)

These values are for zero allowable damage, units randomly placed in layers two units thick, and minor or no wave overtopping.

A tetrapod consists of four tapered legs extending outward from a common point at approximately equal angles to each other; a tribar has three parallel circular cylinders connected by a Y-shaped member that connects to the center point of each cylinder and is normal to the axes of the three cylinders; and a dolos is like the letter H, with the vertical legs rotated 90° to each other (see U.S. Army Coastal Engineering Research Center, 1973 for specifications and information on other shapes in use).

Note the significant effect of unit shape on the stability coefficient, which is inversely proportional to the armor unit weight required. The stability coefficients given for riprap are for the weight of the median stone size in a gradation from $0.22W$–$3.6W$.

The stability coefficient can be increased if a certain percent of damage is allowed. The allowable increase is demonstrated by Table 6.3 which gives stability coefficient values as a function of percent damage for rough angular quarrystone, tetrapods, and tribars. Percent damage is defined as the percent volume of units displaced from a zone extending from the middle of the breakwater crest down the seaward slope to a depth of one wave height below the still water level.

TABLE 6.3. K_D AS A FUNCTION OF PERCENT DAMAGE

Armor Unit	Percent Damage					
	5–10	10–15	15–20	20–30	30–40	40–50
Rough, angular quarrystone	4.9	6.6	8.0	10.0	12.4	15.0
Tetrapods	10.8	13.4	15.9	19.2	23.4	27.8
Tribars	14.2	19.4	26.2	35.2	41.8	45.9

From: U.S. Army Coastal Engineering Research Center, 1973. For breakwater trunk, two random placed layers, nonbreaking waves, minor or no overtopping.

6.6. RIGID VERTICAL WALLS

Several classes of coastal structures including concrete and steel sheet pile bulkheads, quay walls, and breakwaters have a rigid vertical (or near vertical) face that must withstand wave attack. If the water is sufficiently deep at the toe of the structure (i.e. $d > 1.3H$) incident waves will reflect from the structure forming a standing wave system with fluctuating pressures. For shallower water depths (i.e. $d < 1.3H$) the waves will break on the structure or they will break seaward of the structure and send a turbulent mass of water forward to impinge on the structure. Each of these conditions (nonbreaking, breaking, and broken waves) will be considered below.

Nonbreaking-wave Forces

Eq. 2.47 gives the pressure in a standing wave as

$$p = -\rho g y + \frac{\rho g H \cosh k(d+y)}{\cosh kd} \cos kx \cos \sigma t \qquad (2.47)$$

where the sum of the normally incident plus reflected wave heights is $2H$. The first term on the right is the hydrostatic pressure and the second term is the dynamic pressure. If the standing wave is formed by complete reflection from a vertical structure (at $x=0$), the dynamic pressure on the structure is

$$p_d = \frac{\rho g H \cosh k(d+y)}{\cosh kd} \cos \sigma t \qquad (6.12)$$

Thus the dynamic pressure at the toe of the structure ($d=-y$) varies between

$$\pm \frac{\rho g H}{\cosh kd}$$

as the standing wave phase at the structure varies from crest to trough. The dynamic pressure distribution given by Eq. 6.12 is shown in Fig. 6.8 for the wave crest and trough phases. These pressure distributions are almost linear from the instantaneous water surface to the bottom and are usually assumed to be linear (see Fig. 6.8 and Prob. 9) for simplified, conservative calculations of standing wave forces and moments on a vertical structure.

Finite wave amplitude affects cause the mean water level at the structure to rise by an amount Δy where (see Miche, 1944)

$$\Delta y = \frac{\pi H^2}{L} \coth kd \qquad (6.13)$$

Also, wave reflection may not be complete owing to energy dissipation at the face of the structure (i.e. $C_R = H_r/H_i < 1.0$). Thus

$$2H = H_i + H_r = (1 + C_R)H_i$$

or

$$H = \frac{(1 + C_R)H_i}{2} \qquad (6.14)$$

where $2H$ is the standing wave height at the structure.

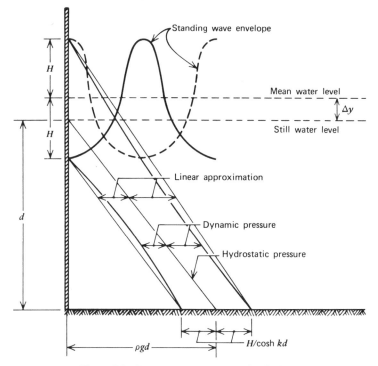

Figure 6.8 Standing wave pressure distributions.

Some walls have failed by falling in the seaward direction. This is probably the result of the hydrostatic or active soil pressure on the leeward side exceeding the minimum force that occurs when a wave trough is acting on the wall.

If $H + \Delta y$ exceeds the elevation of the wall crest above the still water level, the usual design procedure is to assume the pressure up to the crest elevation is the same as if the wall had extended up to $H + \Delta y$. This assumption is probably conservative as flow over the wall crest relieves the pressure distribution on the wall just below the crest.

Breaking-wave Forces

When a wave breaks directly on a vertical structure there is a dynamic impact force on the structure acting around the still water line. This is superimposed on the normal hydrostatic force. Often, a pocket of air is trapped beneath the jet of the breaking wave crest which, through compression of the air, cushions the impact of the breaking waves and reduces the impact pressure. However, on rare occasions, an extremely high intensity, short duration (0.001–0.01 sec) pressure can be exerted by a breaking wave when it breaks just at the wall and traps a thin lense of air. Bagnold (1939) and Denny (1951) report results of these high intensity pressures and discuss possible reasons for their occurrence.

A commonly used design procedure for breaking wave forces presented by Minikin (1950) is based on the data collected by Bagnold (1939). Minikin's method defines the breaking wave impact pressure as being symmetrical with a maximum pressure p_m at the mean water level that decreases parabolically to zero at $\pm H_b/2$, where H_b is the breaking wave height. Thus the impact force per unit length of structure is $p_m H_b/3$. For the maximum pressure, Minikin presents the empirical equation

$$p_m = \frac{2\pi\gamma H_b d\,(d'+d)}{L'd'} \tag{6.15}$$

where d is the water depth at the toe of the structure and d' and L' are the water depth and wave length, respectively, one wave length seaward of the structure. The hydrostatic pressure distribution owing to a water depth $d+H_b/2$ is added to allow calculation of the total breaking wave force and moment.

Carr (1954) has also collected experimental data on forces caused by waves breaking on vertical and inclined walls. His data (which are also presented in Wiegel, 1964) are plotted in terms of the dimensionless maximum force and moment arm versus wave steepness, relative water depth, and offshore slope. While the method from Minikin and Carr's data are useful for design, the subject of breaking wave forces on a variety of structure shapes is in need of much additional research.

Broken Wave Forces

When waves break completely seaward of a structure, the structure can be subjected to a surge of water that exerts a dynamic pressure. The author knows of no experimental data or well-developed theories for predicting the forces on a vertical wall due to broken waves. The U.S. Army Coastal Engineering Research Center (1973) presents a method (believed to be conservative) for broken wave force prediction that is based on a number of simplifying assumptions. One assumption is that after a wave breaks the water mass surges forward with an initial velocity that is the same as the wave celerity before breaking (i.e. $\sqrt{gd_b}$). It is then assumed that the water mass velocity and vertical thickness decrease linearly to zero at the hypothetical point of maximum runup (hypothetical because the surge is interrupted by the structure). The kinetic energy of the surging water mass is converted to dynamic pressure at the structure to produce the resulting dynamic force, which is added to the hydrostatic force to predict the resulting force and moment. This method gives an "order of magnitude" estimate of the broken wave force but model tests are recommended if more accurate force estimates are desired.

6.7. OTHER FORCES ON COASTAL STRUCTURES

At some nearshore locations other forces besides those caused directly by wave and current activity can be significant to the design of structures. Wind forces, particularly in regions such as the hurricane belt where extreme wind velocities occur, can have an important effect on structures that extend above the sea surface. In cold regions ice (often driven by wind and currents) can cause a variety of loadings on a structure. Along the Pacific coast of the United States and other seismically active coastal regions, earthquake induced vibrations can be structurally damaging. Each of these will be briefly discussed in this section.

Wind Forces

Section 4.10 discusses the selection of a design wind speed, direction, and duration for hurricanes and other major storms. Besides these wind characteristics, information is needed on the vertical wind velocity profile near sea level and on the level of gustiness of the wind. The velocity profile is given by the classic Prandtl-Karman velocity distribution law for smooth and rough surfaces. The law allows prediction of the time-averaged velocity at a given elevation above the water surface (see Bretschneider, 1969). Owing to wind gusts, the average wind speed over a short duration (e.g. few seconds) can greatly exceed the 1- or 10-min average speed. Measurements taken in Australia (see Bretschneider, 1969) showed probable values of, for example, the 5-sec average speed to be 116 knots, when the 1-min and 1-hr average speeds were 99 and 80 knots, respectively. Depending on the structure size and the wind speed, a 5-sec gust might be large enough to envelope a structure and cause damage.

The wind force in a structure is given by

$$F = \frac{C_D}{2} \rho_f A U^2 + C_M \mathcal{V} \rho_f \frac{dU}{dt} \qquad (6.4)$$

where drag coefficient values are given by Hoerner (1965) and others as mentioned in Section 6.1. The inertia term in Eq. 6.4 is usually neglected in wind force calculations. However, it can be significant in some instances, for example, when extreme gustiness occurs. Although water is about 800 times as dense as air, the squared wind velocity can possibly exceed the squared water particle velocity in a wave or current by a factor of 800, causing wind forces of equal magnitude to wave or current forces. Also, water entrained in air near sea level will increase the air density.

As is the case with submerged structures, vortex shedding as wind blows past a structure can cause a resonant response and related structural problems. For a further discussion of this and other wind loading phenomena in the coastal zone, Bretschneider (1969) and the Task Committee on Wind Forces (1961) are recommended.

Ice Forces

In cold regions ice can have a major impact on the design of coastal structures, planning and operation of harbors and navigation, and beach processes. Information on sea ice properties and ice effects in the coastal zone can be found in the Proceedings of the biannual International Conferences on Port and Ocean Engineering under Artic Conditions, held since 1971, and in the general articles by Peyton (1968).

The tensile and compressive strength of sea ice is quite variable and is dependent on salinity, temperature, depth within the ice sheet, ice growth rate, and the rate at which a load is applied to the ice. In addition to a knowledge of the strength of sea ice, information on such factors as the expected return period for given ice thicknesses; lateral extent of ice floes that commonly occur; tidal range; and expected speeds and directions of ice floe movement as the ice is driven by wind and tidal currents, is needed for economical designs. Selection of design ice properties is best based on empirical data from a field sampling program at the proposed project site conducted over a period of as many winters as practicable.

There are several ways in which ice can exert forces on structural members including:

1. Moving sheet ice driven by wind and/or currents will exert a horizontal force on a structure at the water line. The force may be caused by the initial impact of the ice or the cutting of a slot through the ice sheet as the ice is crushed by the structure. Ice sheets can be as much as a meter or more thick and, when being crushed, can exert pressures on the order of 200–300 n/cm^2 of frontal projected area. There is evidence (Peyton, 1968) that the peak crushing strength of sheet ice occurs at extremely low (virtually zero) ice speeds rather than at the more visually dramatic

higher speeds. The nature of ice failure by crushing is such that the structural loading is often cyclic, with a rate of a cycle/sec or faster.

2. Inclined structural members will lift the ice sheet and cause ice failure by bending, which results in a much smaller ice force than failure by crushing. Structural members such as piles can often be designed with a conical section over the tidal range to cause bending failure of the ice.

3. Ice frozen to a structure at high tide can exert a significant vertical load as the tide drops and similarly, ice frozen to a structure at low tide can exert a buoyant force as the water level rises. During a thaw, large ice blocks frozen to structural members can move rapidly (e.g., slide down a pile) and cause damage.

4. Ice sheet edges resting on riprap and moving with the tide or currents can pluck riprap units to seriously degrade the revetment.

5. Damage can be caused by freezing and expansion of seawater in cracks and other small openings of structural members.

Much remains to be learned about the various types of ice forces on structures but, particularly with the expanded offshore oil exploration programs in northern latitudes, an increased effort to understand these phenomena is underway.

Earthquake Loadings

Besides the damage caused by earthquake-generated tsunamis discussed in Section 4.4, earthquakes can cause direct coastal damage in a variety of ways. A map of recent earthquake epicenter locations (e.g. see p. 23, Wiegel 1970) will show that most earthquakes occur in coastal regions, particularly around the entire rim of the Pacific Ocean. This subsection will very briefly discuss some of the modes by which earthquakes can affect coastal structures. A more thorough discussion of these and related matters is presented in Wiegel, 1970.

Direct ground shaking will cause major structural excitation over a broad region (tens of kilometers wide) surrounding the earthquake epicenter. Earthquakes having a Richter scale magnitude greater than 5 will probably generate ground motion sufficiently severe to damage some structures. The spectrum of shaking motions is commonly like "white noise," with a range of periods from a fraction of a second to 5 sec or higher and a maximum acceleration that may exceed 0.5 g. The shaking may last up to a few minutes, but the critical high intensity phase is usually shorter (5–15 sec) with higher frequency (<0.5-sec period) oscillations. Chapter 5 of Wiegel, 1970, suggests excitation spectra for the design of earthquake-prone structures. The design of partially or totally submerged structures is complicated by the interaction of structural vibrations with the surrounding water through hydrodynamic drag and inertia forces, which serve to dampen structure response.

Near the epicenter, fault displacement can cause uplift or subsidence of the earth, which can have a major impact on coastal structures that survive the shaking forces. The 1964 Alaskan earthquake caused uplift of 2 m at Cordova, Alaska, which reduced the water depths in a small boat basin from 4 m to less than 2 m (Arno, 1965). Little structural damage was done to the basin rubble mound breakwater and appurtenances.

Shaking of the ground can also affect coastal structures in less direct ways. Underwater and shoreline landslides may be generated. Also, seismic waves may cause compaction of cohesionless soils resulting in land settlement. Loose saturated cohesionless soils (possibly retained land fills), which may be common in some coastal regions, can become liquified to form a quick condition resulting in the sinking or overturning of structures.

6.8. SUMMARY

A survey of the various types of wave-structure interactions that occur in the coastal zone has been presented. Wave forces on structures and the effects of structures on wave motion must be adequately quantified for the structural and functional design of coastal structures. The functional design of coastal structures also strongly depends on their interaction with coastal sediment transport processes. Coastal zone transport processes and the effect of coastal structures are presented in Chapter 7.

6.9. REFERENCES

Aagard, P. M. and R. G. Dean (1969), "Wave Forces: Data Analysis and Engineering Calculation Method," *Proceedings, Offshore Technology Conference*, Houston, paper 1008.

Agerschou, H. A. and J. J. Edens (1965), "Fifth and First Order Wave-force Coefficients for Cylindrical Piles," *Proceedings, Santa Barbara Coastal Engineering Specialty Conference*, American Society of Civil Engineers, pp. 219–248.

Ahrens, J. P. (1975), "Large Wave Tank Tests of Riprap Stability," Tech. Memo 51, U.S. Army Coastal Engineering Research Center, Ft. Belvoir, Virginia, 41 p.

Arno, N. (1965), "Restoring a Small Boat Basin Damaged by the 1964 Alaska Earthquake," *Proceedings, Santa Barbara Coastal Engineering Specialty Conference*, American Society of Civil Engineers, pp. 861–888.

Bagnold, R. A. (1939), "Interim Report on Wave Pressure Research," *Journal, Institution of Civil Engineers*, London, pp. 202–226.

Beattie, J. F., L. P. Brown and B. Webb (1971), "Lift and Drag Forces on a Submerged Circular Cylinder," *Proceedings, Offshore Technology Conference*, Houston, paper 1358.

Blumberg, R. and A. M. Rigg (1961), "Hydrodynamic Drag at Supercritical Reynolds Numbers," presented at ASME meeting, Los Angeles, June, 18 p.

Brater, E. F. and R. Wallace (1972), "Wave Forces on Submerged Pipelines," *Proceedings, Thirteenth Conference on Coastal Engineering*, Vancouver, pp. 1703–1722.

Bretschneider, C. L. (1969), "Overwater Wind and Wind Forces," *Handbook of Ocean and Underwater Engineering*, J. J. Meyers, Editor, McGraw-Hill, New York, pp. 12.2–24.

Brown, R. J. (1967), "Hydrodynamic Forces on a Submarine Pipeline," *Journal Pipeline Division, American Society of Civil Engineers*, March, pp. 9–19.

Carr, J. H. (1954), "Breaking Wave Forces on Plane Barriers," Cal. Tech. Hydromechanics Lab Rept. E-11.3.

Chamberlain, R. S. (1970), "Undersea Oil Storage Tank," *Civil Engineering*, August, pp. 57–60.

Chang, P. Y. and W. D. Pilkey (1971), "The Analysis of Mooring Lines," *Proceedings, Offshore Technology Conference*, Houston, paper 1502.

Chakrabarti, S. K. (1973), "Wave Forces on Submerged Objects of Symmetry," *Journal, Waterways, Harbors and Coastal Engineering Division*, American Society of Civil Engineers, May, pp. 145–164.

Chakrabarti, S. K. and W. A. Tam (1973), "Gross and Local Wave Loads on a Large Vertical Cylinder—Theory and Experiment," *Proceedings, Offshore Technology Conference*, Houston, paper 1818.

Chen, K. and R. L. Wiegel (1970), "Floating Breakwater for Reservoir Marinas," *Proceedings, Twelfth Conference on Coastal Engineering*, Washington, D.C., pp. 1645–1665.

Crooke, R. C. (1955), "Re-analysis of Existing Wave Force Data on Model Piles," Tech. Memo 71, U.S. Army Beach Erosion Board, Washington, D.C., 21 p.

Cross, R. H. and R. L. Wiegel (1965), "Wave Forces on Piles: Tables and Graphs," Report HEL 9-5, University of California, Berkeley, 57 p.

Denny, D. F. (1951), "Further Experiments on Wave Pressures," *Journal, Institution of Civil Engineers*, London, pp. 330–345.

Dominguez, R. F. and R. W. Filmer (1971), "Discrete Parameter Analysis as a Practical Means of Solving Mooring Behavior Problems," *Proceedings, Offshore Technology Conference*, Houston, paper 1505.

Evans, D. J. (1969), "Analysis of Wave Force Data," *Proceedings, Offshore Technology Conference*, Houston, paper 1005.

Garrison, C. J. and P. Y. Chow (1972), "Wave Forces on Submerged Bodies," *Journal, Waterways, Harbors and Coastal Engineering Division*, American Society of Civil Engineers, August, pp. 375–392.

Garrison, C. J. and V. S. Rao (1971), "Interaction of Waves with Submerged Objects," *Journal, Waterways, Harbors and Coastal Engineering Division*, American Society of Civil Engineers, May, pp. 259–277.

Grace, R. A. (1971), "The Effects of Clearance and Orientation on Wave Induced Forces on Pipelines," Look Lab Report 15, University of Hawaii, 53 p.

Harleman, D. R. F. and W. C. Shapiro (1955), "Experimental and Analytical Studies of Wave Forces on Offshore Structures," Tech. Rept. 19, Hydromechanics Lab, MIT.

Harleman, D. R. F. and W. C. Shapiro (1961), "Investigation on the Dynamics of Moored Structures in Waves" *Proceedings, Seventh Conference on Coastal Engineering*, The Hague, pp. 746–765.

Helfinstine, R. A. and J. W. Shupe (1972), "Lift and Drag on a Model Offshore Pipeline," *Proceedings, Offshore Technology Conference*, Houston, paper 1568.

Herbich, J. B. and G. E. Shank (1970), "Forces Due to Waves on Submerged Structures: Theory and Experiment," *Proceedings, Offshore Technology Conference*, Houston, paper 1245.

Hoerner, S. F. (1965), *Fluid-Dynamic Drag*, published by the author, 430 p.

Hudson, R. L. (1959), "Laboratory Investigation of Rubble-Mound Breakwaters," *Journal, Waterways and Harbors Division*, American Society of Civil Engineers, September, pp. 93–121.

Ippen, A. T. (1966), *Estuary and Coastline Hydrodynamics*, McGraw-Hill, New York, 744 p.

Jen, Y. (1968), "Laboratory Study of Inertia Forces on a Pile," *Journal, Waterways and Harbors Division*, American Society of Civil Engineers, February, pp. 59–76.

Keim, S. R. (1956), "Fluid Resistance to Cylinders in Accelerated Motion," *Journal, Hydraulics Division*, American Society of Civil Engineers, December, paper 1113.

Keulegan, G. H. and L. H. Carpenter (1958), "Forces on Cylinders and Plates in an Oscillating Fluid," *Journal, Research, National Bureau of Standards*, Vol. 60, paper 2857, pp. 423–440.

Kincaid, G. A. (1960), "Effects of Natural Period Upon the Characteristics of a Moored Floating Breakwater," Thesis, Dept. of Civil and Sanitary Engineering, MIT, Cambridge.

Kowalski, T. (1974), Floating Breakwaters Conference Papers, University of Rhode Island, Marine Technical Report No. 24, Kingston, 304 p.

Laird, A. D. K, C. A. Johnson and R. W. Walker (1959), "Water Forces on Accelerated Cylinders," *Journal, Waterways and Harbors Division*, American Society of Civil Engineers, August, pp. 99–119.

MacCamy, R. C. and R. A. Fuchs (1954), "Wave Forces on Piles: A Diffraction Theory," Tech. Memo 69, U.S. Army Beach Erosion Board, Washington, D.C., 17 p.

Miche, M. R. (1944) "Mouvements Ondulatoires de la Mer en Profondeur Constante ou Decroissante," translation in University of California Wave Research Lab Report 3-363, 1954, pp. 96.

Minikin, R. R. (1950), *Winds, Waves and Maritime Structures*, Charles Griffen and Co., London.

Morison, J. R., J. W. Johnson, M. P. O'Brien and S. A. Schaaf (1950), "The Forces Exerted by Surface Waves on Piles," *Petroleum Transactions, American Institute of Mining Engineers*, Vol. 189, pp. 145–154.

Ouellet, Y. (1972), "Effect of Irregular Wave Trains on Rubble Mound Breakwaters," *Journal, Waterways, Harbors and Coastal Engineering Division*, American Society of Civil Engineers, February, pp. 1–14.

Paape, A. and H. N. C. Breusers, "The Influence of Pile Dimensions on Forces Exerted by Waves," *Proceedings, Tenth Conference on Coastal Engineering*, Tokyo, pp. 840–849.

Peyton, H. R. (1968), "Ice and Marine Structures," *Ocean Industry*, Houston, 3 parts: March, September, December.

Ried, R. O. (1958), "Correlation of Water Level Variations with Wave Forces on a Vertical Pile for Nonperiodic Waves," *Proceedings, Sixth Conference on Coastal Engineering*, Council on Wave Research, pp. 749–786.

Rogan, A. J. (1969), "Destruction Criteria for Rubble Mound Breakwaters," *Proceedings, Eleventh Conference on Coastal Engineering*, London, pp. 761–788.

Sarpkaya, T. and C. J. Garrison (1963), "Vortex Formation and Resistance in Unsteady Flow," *Journal, Applied Mechanics*, Vol. 30, March, pp. 16–24.

Seymour, R. J. and J. D. Isaacs (1974), "Tethered Float Breakwaters," *Floating Breakwaters Conference*, University of Rhode Island, Marine Technical Report No. 24, Kingston, pp. 55–72.

Task Committee on Wind Forces (1961), "Wind Forces on Structures," *Transactions*, American Society of Civil Engineers, Part II, pp. 1124–1198.

U.S. Army Coastal Engineering Research Center (1973), *Shore Protection Manual*, 3 Vols., Government Printing Office, Washington, D.C.

Wiegel, R. L. (1964), *Oceanographical Engineering*, Prentice-Hall, Englewood Cliffs, New Jersey, 532 p.

Wiegel, R. L. (1970), Coordinating Editor, *Earthquake Engineering*, Prentice-Hall, Englewood Cliffs, New Jersey, 518 p.

Wiegel, R. L., K. E. Beebe and J. Moon (1957), "Ocean Wave Forces on Circular Cylindrical Piles," *Journal, Hydraulics Division*, American Society of Civil Engineers, April, paper 1199.

Wiegel, R. L. and R. C. Delmonte (1972), "Wave-induced Eddies and Lift Forces on Circular Cylinders," Report HEL 9-19, University of California, Berkeley, 39 p.

Wilson, B. (1965), "Analysis of Wave Forces on a 30-inch Diameter Pile Under Confused Sea Conditions," Tech. Memo 15, U.S. Army Coastal Engineering Research Center, Washington, D.C., 85 p.

Wilson, B. W. (1967), "Elastic Characteristics of Moorings," *Journal, Waterways and Harbors Division*, American Society of Civil Engineers, November, pp. 27–56.

Wilson, B. W. and D. H. Garbaccio (1967), "Dynamics of Ship Anchor-Lines in Waves and Current," *Proceedings, Civil Engineering in the Oceans Conference*, San Francisco, pp. 277–304.

Wilson, B. W. and R. O. Reid (1963), discussion of "Wave Force Coefficients for Offshore Pipelines," *Journal, Waterways and Harbors Division*, American Society of Civil Engineers, February, pp. 61–65.

6.10. PROBLEMS

1. A vertical 1-m diameter circular pile standing in water 30 m deep is subjected to a 4-m high, 11-sec wave. Calculate and plot the drag, inertia, and total force distributions along the pile at the instant the wave crest is 20 m seaward of the pile.

2. Consider the pile in Prob. 1. Calculate and plot the ratio of total drag to total inertia force on the pile for an 11-sec wave as a function of wave height for heights ranging from 0–10 m. Assume the wave crest is one-eighth of a wave length seaward of the pile, and let $C_D = 1.0$ and $C_M = 1.5$.

3. A horizontal cylindrical cross brace on an offshore tower having a 0.8-m diameter and a length of 9 m is 6 m below the still water level. The water depth is 30 m. For a 12-sec, 5-m high wave approaching normal to the axis of the brace, calculate and plot the drag, inertia and total force on the brace as a function of time for one wave period.

4. Demonstrate that $u_m T/D = 2\pi\zeta/D$, relate $2\pi\zeta/D$ to H/D, and comment on the physical significance of these parameters to the force exerted by waves on a pile.

5. A 0.3-m diameter vertical pile supporting a pier is situated in water 8 m deep. The pier is at a site on a deep lake where the fetch is 24 miles and the design wind speed is 60 knots with a duration of 2 hr. Determine the design moment around the mudline due to wave forces and discuss the basis for the design wave you selected.

6. A 0.8-m diameter submerged pipeline resting on a horizontal bottom (assume $\mu = 0.8$) is subjected to a 4-m high, 13-sec wave propagating normal to the pipeline axis. The water depth is 14 m. What minimum weight per meter should the pipe have?

7. A revetment similar to that shown in Fig. 6.7 is placed on the face of a small earth dam (1:3 slope). The toe of the revetment is at a depth of 2 m below the design water level and the bottom slope in front of the revetment is 1:50. For a design wave at the structure having $H_s = 1.9$ m and $T_s = 4.5$ sec determine the median stone size required for zero damage and select a desired crest elevation for the revetment.

8. What weight concrete tribars are needed as armor units for a breakwater (cross-section similar to Fig. 6.6) if no damage is allowed? The water depth at the seaward toe is 5 m and the bottom offshore of the breakwater has a 1:10 slope. A design wave with $H_s = 3.1$ m and $T_s = 8$ sec (no refraction or diffraction on shoaling) in deep water is to be used.

9. By integrating the pressure over a vertical wall from the bottom $-d$ to the water surface η, derive an equation for the force due to a standing wave as a function of time. Compare the result obtained from this equation with that calculated using the linear pressure distribution assumption for the force under a wave crest exerted by a 1-m, 6-sec wave on a vertical wall with a toe depth of 3 m.

10. A rigid vertical wall has a reflection coefficient estimated to be 0.8. The water depth at the toe is 4.2 m. What is the moment around the toe exerted by a 2-m, 6-sec wave?

11. A vertical wall in water 2 m deep is fronted by a 1:10 slope. Determine the height of a 7-sec. wave that will break at the wall and determine the total force exerted by this breaking wave per foot of wall.

12. A vertical wall is constructed on a beach with a 1:15 slope. The toe of the wall is at an elevation of $+0.5$ m above mean water level. For a normally incident 2-m, 8-sec. wave, estimate the dynamic pressure on the wall using the approach outlined in Section 6.6c. If the dynamic pressure is assumed to vary uniformly over the wall, what is the dynamic force on the wall?

Seven

COASTAL ZONE PROCESSES

The zone of interest in this chapter is that section of the coast located between the offshore point where shoaling waves begin to move sediment and the onshore limit of active marine processes. The latter is usually delineated by a dune field or cliff line. Some shorelines consist of long sand or gravel beaches occasionally interrupted by a stream, tidal inlet or rocky headland. Others consist of wave-resistant cliffs near the water's edge with an occasional small pocket beach and a thin layer of sediment overlying bedrock in the shallow nearshore waters. The emphasis in this chapter is on those sections of the coast having a long sandy beach.

Beach and shallow water sediments are continuously responding to direct wave action, wave-induced littoral currents, wind, and tidal currents. The stability of a section of sedimentary shoreline depends on a balance between the volume of sediment available to that section and the net onshore-offshore and alongshore sediment transport capacity of waves, wind, and currents in that section. The shoreline may thus be eroding, accreting or remaining in equilibrium. If equilibrium does exist, it is at best a "dynamic equilibrium," where the shoreline is responding to continuously variable winds, waves, and currents. Also, the supply of sediment is usually irregular in time and space. Dynamic equilibrium usually means that the average shoreline position is relatively stable over a period of months or years while the instantaneous position undergoes short-term oscillations.

Construction of structures, dredging, beach nourishment with sand, and other developments in the coastal zone have often been carried out to limit or reverse shoreline erosion or accretion. At times, these coastal developments have upset an existing dynamic equilibrium of the shoreline. The result is a continuing shoreline change or a new equilibrium condition which may or may not be desirable. Coastal developments can affect coastal zone processes by (1) changing the rate and/or characteristics of sediment supplied to the coast, (2) adjusting the level of wave energy flux to the coast, and (3) directly interfering with coastal sediment transport processes. Two examples of the first of these are (1) construction of a dam that traps sediment on a river discharging on the updrift coast, thus depriving the shoreline of an adequate supply of sediment, and (2) periodic placement of sand directly on a beach to nourish it. Examples of the second and third effects respectively are: construction of an offshore breakwater that intercepts waves approaching a beach to, in turn, reduce the wave-induced longshore sediment transport, and construction of a groin across the surf zone to directly interrupt wave-induced longshore currents and sediment transport.

This chapter discusses the important coastal sediment properties and active hydrodynamic and sedimentary processes as well as the major effects of the works of man on the coastal zone.

7.1. BEACH SEDIMENT PROPERTIES AND ANALYSIS

Of greatest interest are those physical properties of beach sediments that control their response to wind, wave, and current action and, in turn, are important to the design of engineering works in the coastal zone. Primary among these properties are the representative sediment's particle size and size distribution. Particle shape and specific gravity as well as the bulk properties of permeability and specific weight will also be considered. The response of a particle to hydrodynamic forces is indicated by the particle-settling or fall velocity in water, a parameter experiencing increased use in studies of beach sediment mechanics. Fall velocity, however, is dependent on the basic properties of size, shape, and specific gravity.

Care must be exercised in collecting beach sediment samples for analysis. The sampling procedure must provide a sufficient volume of sample for analysis; must not disturb the sediment properties of interest (e.g. loss of fines as sample is retrieved); and must provide samples that are spatially and temporally representative of those characteristics of the beach environment important to the problem at hand. Samples are best taken with a cylindrical core pressed at least a few centimeters into the beach face or sea bed. A variety of sampling devices have been developed for underwater sampling.

Krumbein (1954) and Krumbein and Slack (1956) discuss several schemes for establishing the number and locations of samples to be taken in a beach sampling program, as well as the statistical reliability of each scheme. In general, samples should be selected from several points across a number of beach profiles lying normal to shore and spaced evenly along the area of interest. The samples selected from each profile should be representative of the different zones crossed by the profile from the dunes to a point seaward of breaking waves. Seasonal variations in beach characteristics must also be considered. Naturally, the sampling program must reflect the intended use of the sample analysis data and the level of funding available for sampling and analysis.

Size and Size Distribution

A full range of sediment sizes from clay (less than one micron diameter) to gravel and boulders (several centimeters diameter) can be found in coastal areas. Most beaches consist of sand with particle diameters between 0.1–1.0 mm. Several formal sediment size classifications have been proposed during the past century; the most commonly used classification being that proposed by Wentworth (1922) and given in Table 7.1. Since most sediment sample size distributions will be skewed with a preponderance of finer sizes, the Wentworth classification was made logarithmic with base 2. This allows a better definition of the finer sizes.

In much of the literature, sediment particle diameters are defined by phi units ϕ proposed by Krumbein (1936) and based on the Wentworth classification. For a particle diameter d in mm,

$$\phi = -\log_2 d \tag{7.1}$$

where the minus sign is used so the more common sediment sizes ($d < 1$ mm) will have a positive phi value. Wentworth size class boundaries are thus whole numbers in phi units. However, the phi unit system can cause some confusion as phi units increase with decreasing sediment size and each whole number interval represents a different range of particle sizes.

The particle size distribution for a sediment sample is best defined by standard statistical techniques. A sample can be represented by a frequency histogram, which is a plot of percent of sample by volume or weight versus particle diameter in millimeters or phi units. The integral of the frequency histogram (cumulative size frequency distribution), which is a plot of cumulative percent coarser or finer than a particular diameter versus particle diameter, is also useful. Griffiths (1967) presents a useful discussion of statistical techniques with particular reference to sediment size analysis.

TABLE 7.1. WENTWORTH SIZE CLASSIFICATION

Class			Particle Diameter	
			mm	phi units
Boulder				
			256	−8
Cobble			128	−7
		Large	64	−6
		Medium	32	−5
Pebble		Small	16	−4
		Very Small	8	−3
			4	−2
		Granule	2	−1
		Very Coarse	1	0
		Coarse		
Sand		Medium	1/2	1
		Fine	1/4	2
		Very Fine	1/8	3
		Coarse	1/16	4
		Medium	1/32	5
Silt		Fine	1/64	6
		Very Fine	1/128	7
		Coarse	1/256	8
		Medium	1/512	9
Clay		Fine	1/1024	10
		Very Fine	1/2048	11
			1/4096	12

The important features of a sediment sample size-frequency distribution can be defined by three parameters: the central tendency (mean, median, or modal diameter); the dispersion or sorting (standard deviation); and the asymmetry (skewness). The median diameter is easier to determine than the mean or modal diameter and it is affected less by extremely large or small sizes than is the mean. As sediment is transported down a river and deposited on a beach, coarser sizes often remain in the river bed and fine sizes are deposited offshore so beach sediments are usually well sorted sand (i.e. have a narrow size range). As discussed above, beach sands, like other sediments, are usually skewed toward finer sizes.

Several descriptive measures, based on sediment size analyses, have been proposed to define the central tendency, sorting, and asymmetry of a sample (see Griffith, 1967, for a summary). One system commonly used in engineering practice was developed by Inman (1952). A portion of this system is presented in Table 7.2 where ϕ_{16}, ϕ_{50}, and ϕ_{84} are the 16, 50, and 84 percent coarser phi diameters from a cumulative size-frequency distribution.

The descriptive measures from Table 7.2 are demonstrated in Fig. 7.1 for a typical sand sample size analysis plotted on log-normal distribution paper. Experience has shown that many beach sand samples have approximately a log-normal distribution and thus should plot close to a straight line on log-normal paper.

In Fig. 7.1 the phi median diameter is 2.13ϕ, which is equivalent to a median diameter, d_{50}, of 0.23 mm. The phi mean diameter 1.81ϕ is a somewhat arbitrarily defined mean having no direct correspondence to the arithmetic mean diameter of the sample.

In an arithmetic normal distribution, the spacing between the 16th and 84th percentiles is two standard deviations. This suggests the definition of the phi deviation measure, which indicates the spread or sorting of the sample from the median size. If the sample were symmetrical (unskewed) it would plot as a straight line in Fig. 7.1 and would be completely defined by the phi median diameter and the phi deviation measure. The phi skewness measure gives an indication of the displacement of the median from the mean and thus indicates the skewness. The phi deviation and phi skewness measures should not be converted to equivalent mm diameters, as these would have no physical meaning.

Sediment Size Analysis

A rough estimate of the mean or median sand particle diameter—useful for quick analysis in the field—can be made by visually comparing field samples with standardized samples of selected sizes. The standardized samples can be glued to a card or kept in small transparent vials. The two most common laboratory methods for more thorough sand size analysis are by sieving and settling tube.

TABLE 7.2. SEDIMENT SAMPLE DESCRIPTIVE MEASURES (AFTER INMAN, 1952)

Measure	Name	Definition
Central tendency	Phi median diameter	$M_{d\phi} = \phi_{50}$
	Phi mean diameter	$M_{\phi} = \frac{1}{2}(\phi_{16} + \phi_{84})$
Sorting	Phi deviation measure	$\sigma_{\phi} = \frac{1}{2}(\phi_{84} - \phi_{16})$
Skewness	Phi skewness measure	$\alpha_{\phi} = \dfrac{M_{\phi} - M_{d\phi}}{\sigma_{\phi}}$

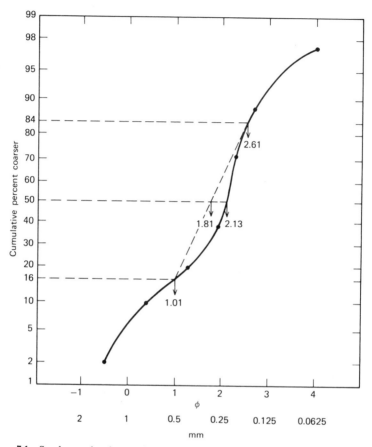

Figure 7.1 Sand sample size analysis; Inman's descriptive measures. $M_{d\phi} = 2.13$, $M_\phi = \frac{1}{2}(2.61 + 1.01) = 1.81$, $\sigma_\phi = \frac{1}{2}(2.61 - 1.01) = 0.80$, $\alpha_\phi = \dfrac{1.81 - 2.13}{0.80} = 0.40$.

Sieving is generally considered the most reliable method of sand size analysis. U.S. Standard Sieve opening sizes vary from 125 mm down to 0.038 mm, with opening sizes below 5.6 mm decreasing by the constant ratio of $\sqrt[4]{2}$ (i.e. 1.19) or $1/4\ \phi$ intervals. For typical beach sands, alternate sieves between 2.0 mm and 0.062 mm (i.e. $1/2\ \phi$ intervals) might typically be used to perform a size analysis.

A dry sample (40-60 grams if fine sand, 100-150 grams if coarse) is passed through the series of sieves, and the percent of total cumulative sample weight collected on successively larger sieves is plotted versus sieve opening size to yield a cumulative size-frequency diagram (Prob. 1). Mahlig and

Reed (1972) present a thorough discussion of recommended sieving practices.

The terminal settling velocity of a particle in still water depends on the particle's specific gravity and shape as well as its size. Thus particle size distribution analysis by a settling tube includes the effects of particle shape (as does sieving) and to a lesser extent specific gravity, which is not too variable for most beach sands. Shape and specific gravity influences are not always undesirable, as our interest in sediment size distributions is often related to the response of sediment to hydrodynamic forces. Settling tube size analyses are faster than sieve analyses and do not require as large a sample.

Two of the more common settling tubes are the Visual Accumulation Tube (Colby and Christensen, 1956) and the Rapid Sediment Analyzer (Schlee, 1966). The VAT, for example, is a transparent, water-filled glass tube with a mechanism for quick release of a sediment sample at the top and a device for recording the sediment-water interface position at the bottom, as settling particles accumulate. The rate of particle accumulation is empirically related to the fall velocities (and thus diameters) of the sample particles, so a trace of the sediment-water interface versus time can be converted directly into a cumulative size-frequency distribution curve. The resulting "fall diameter" or "settling diameter" is the diameter of a quartz sphere that would have the same settling velocity as the sand grain in 24°C, still, fresh water. This would not necessarily be equal to the sieve diameters for the same sand grains.

For noncolloidal particles finer than sand, the hydrometer or pipette methods (Krumbein and Pettijohn, 1938), which also involve particle settling analysis, may be used for size determination.

Other Sediment Properties

Besides affecting hydrodynamic response, particle shape may be an indicator of a sediment's source and recent history. Continuous abrasion by agitation in the surf zone tends to increase particle roundness. Shape may be evaluated by observing particles with a magnifying glass and comparing their shape to a chart standard.

Most beach sands consist predominantly of quartz (2.65 specific gravity), with a smaller portion of feldspars (2.54–2.64 s.g.). They may also contain calcite (2.72 s.g.) in the form of shells, and a small percentage of heavy minerals (s.g. > 2.87). The sand composition may occasionally provide an indication of the source region or regions (see Trask, 1952; Kamel and Johnson, 1963). The bulk specific gravity of sand will usually be 1.45–1.85 dry and 1.90–2.15 when saturated with water.

The permeability of the sea bed controls flow in and out of the bed as a wave passes, and thus affects the rate of wave energy dissipation. If a beach face is very permeable, the return flow from wave runup will seep into the face rather than return to the sea by flowing over the surface. This, in turn, will affect wave runup and resulting beach face slope. From experiments on sand, Krumbein and Monk (1942) found the permeability K in darcys to be given by

$$K = 760(d_{50})^2 e^{-1.31\sigma_\phi} \qquad (7.2)$$

Thus beach sand permeability increases with the square of the median grain diameter and with increased sorting.

7.2. BEACH PROFILES

Beach profiles, measured normal to the shoreline over the active zone, are of great importance to coastal engineering studies. This active zone typically extends from the onshore dune or cliff line to a point offshore where there is little significant sediment movement owing to wave action (usually at a depth of about 10 m for the open ocean). Over the active zone, a portion of the beach profile can change drastically in a few hours with a sudden increase in wave activity. Profile data are important to an understanding and quantification of coastal zone processes, as well as to the planning of beach nourishment projects and the functional design of seawalls, piers, groin fields, marine waste outfall pipelines, and other classes of coastal structures. In this section, beach profile response to waves, typical profile recession rates, the related onshore and offshore movement of sand, and the variation of sand size characteristics along a profile will be discussed. In any discussion of beach profiles it is important to remember that the profile characteristics and response can never be fully separated from the influences of longshore sand transport and beach geometry changes.

Figure 7.2 shows typical, somewhat schematic, beach profiles commonly developed by the two wave climate extremes: (1) storm waves of high amplitude and steepness, and (2) calm condition waves of lower amplitude and steepness. High winds may transport large volumes of sand on to the dunes, and waves generated during extreme storms may attack a cliff or dune face causing it to recede, but most of the time, change is limited to that shown by the profiles in Fig. 7.2.

A beach profile typically contains one or two landward sloping berms in the backshore region, a foreshore region of active wave runup, and a

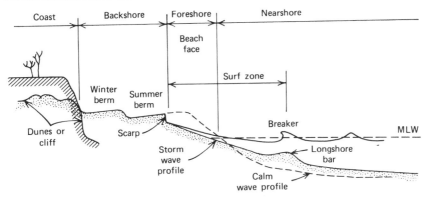

Figure 7.2 Typical beach profiles (vertical scale exaggerated).

concave nearshore submarine profile, possibly having one or more wave breakpoint bars lying approximately parallel to shore. During calm wave conditions sand is slowly moved shoreward to build the beach face in the foreshore zone and to extend the berm, thus causing a steeper beach face profile slope. With the return of storm waves, sand is moved offshore from the foreshore zone and a scarp is cut into the berm. If the storm duration and wave activity are sufficiently great the berms may be completely removed and the beach face profile cut back to the cliff or dunes.

At most coastal locations, storm waves predominate during winter months and calmer waves occur during the summer. Thus, during the winter a storm wave profile with only a winter berm will be common. If the beach is receding there may be no berm during much of the winter and storm waves will then attack the cliff or dune face. During the summer (except after hurricanes or other storms) a calm wave profile with a summer berm will be commonly found on all but the most rapidly receding beaches.

Nearshore bar geometry and spacing closely respond to the predominant waves. With the occurrence of higher waves a bar will move seaward (as will the wave breaking point), and the bar will grow in size. With a return of low waves the bar may be stranded at its seaward position while a new smaller bar is formed closer to shore. During extremely low waves no bars will be built. Bars are also relatively less common where the tide range is great. Bars are usually continuous alongshore for long sections of sandy coast.

Foreshore and nearshore beach profiles are usually measured by standard level and tape or stadia surveying techniques. A boat or amphibious craft with a leadline or echo sounder for depth measurements is used offshore. Offshore positioning can be done by triangulation with a sextant but faster and more accurate positioning can be accomplished with electronic range finders, particularly for complex hydrographic surveys near tidal inlets, harbor entrances, and so on.

Early small-scale laboratory studies indicated that the existence of either a storm or calm wave beach profile could be related to the wave steepness (H_0/L_0). However, this was not completely supported by prototype scale laboratory and field data, as the sediment size characteristics are also important. A better prediction parameter seems to be the fall time parameter F_0 (Dean, 1973) where

$$F_0 = \frac{H_0}{V_f T} \qquad (7.3)$$

and V_f is the terminal settling velocity for the median diameter particle. For $F_0 < 1$ profile accretion occurs and for $F_0 > 2$ erosion is most likely.

A beach profile may recede as much as 30 m or more in the landward direction during a single intense storm. If much of the beach sand is moved too far offshore for subsequent return to the beach face and nearshore zone by calm weather waves, and there is insufficient net sand accumulation from the longshore sand transport, permanent recession of the shoreline will result. These short-term, seasonal, and long-term fluctuations of the beach profile and shoreline position must be documented before a successful design of most coastal structures can be carried out.

Seelig and Sorensen (1973) studied shoreline position changes during the past century at 226 points along the 400-mile Texas coast. The MLW shoreline position at 50 percent of the points showed a small rate of change of less than ± 2 m/year. But at 40 percent of the points a shoreline recession in excess of 2 m/year, with extreme rates in excess of 10 m/year, was observed. The remaining 10 percent of the points, mainly near jettied inlet entrances, showed shoreline advances in excess of 2 m/year. The shoreline change rates observed in Texas are not unlike those in many other United States coastal regions.

The sand size, as indicated by the median diameter of a sample, will vary along a beach profile, particularly for beaches with coarse composite sizes. From a study of samples from several Pacific coast beaches, Bascom (1951) found the sand to be coarsest at the plunge point in the wave breaker zone where the highest turbulence levels occur. The next coarsest sand was found on the berms, perhaps because of the winnowing of finer sizes by the wind. Where dunes exist, the dune sand was progressively finer in the landward direction. Sand sizes also progressively decreased seaward of the surf zone.

There is a general correlation between beach face slope, sand size, and exposure to wave attack. For a given wave climate, the coarser the median sand diameter on the beach face, the steeper the beach face will be. Typical beach face slopes can vary between 1:5 and 1:100. Beach face median sand diameters for United States beaches vary from approximately 1 mm to as low as 0.1 mm, depending on the sand sources and beach exposure. For a given sand size, more exposed beaches will have a flatter beach face slope. Also, the berm crest elevation is approximately equal to the average wave runup at higher tides.

7.3. NEARSHORE CIRCULATION

Currents in the nearshore zone may be generated by sustained winds, river outflow, or tide-generated flow at inlet entrances. However, the most common nearshore currents are those that flow alongshore in the surf zone and are generated by waves breaking at an angle to the shore. These wave-generated longshore currents and the associated surf zone wave activity are responsible for most of the longshore sediment transport in the nearshore zone.

Figure 7.3 is a schematic plan view of a portion of the foreshore-nearshore zone with waves approaching at an angle to the shoreline, breaking, and running up on the beach face. Also shown is the resulting longshore current velocity distribution, which extends from a point just seaward of the wave

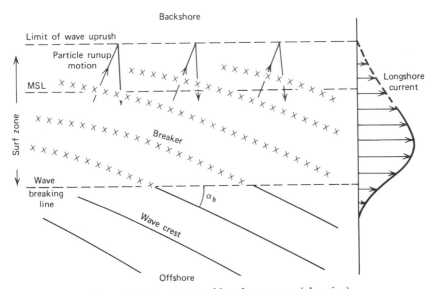

Figure 7.3 Wave-generated longshore current (plan view).

breaking line to the beach face. The maximum longshore current velocity is usually in the surf zone just inside of the breaker line.

Based on numerous field measurements in southern California, Ingle (1966) reported longshore current velocity measurements as high as 1.3 m/sec with the average value for all measurements being 0.3 m/sec. Velocities seaward of the breaker line never exceeded 0.3 m/sec. These values are in general agreement with field data reported by Szuwalski (1970), Komar and Inman (1970), and others.

The mechanism primarily responsible for wave-generated longshore currents is the longshore component of excess momentum flux (or radiation stress) in oblique shoaling waves (Longuet-Higgins, 1970). Also, variations in the longshore distribution of wave breaker heights (owing to refraction, diffraction, etc.) will cause longshore variations in the surf zone wave setup and the generation of longshore currents from areas of high waves to areas of low waves (Komar, 1975). These two mechanisms may support or oppose each other in establishing a longshore current. On most reasonably uniform coastlines, the former mechanism (longshore momentum flux) will predominate.

Successive waves in a wave train will have different heights and periods so the normal and longshore components of momentum flux are time variant. Thus the wave-generated longshore current often exhibits a pulsating behavior with a period in the order of a few minutes.

Galvin (1967) presents a review of pre-1967 attempts to develop analytical equations for longshore current velocity. These were based on energy balances, conservation of momentum flux, or conservation of mass flux. Regression analysis (see Sonu et al. 1967) has also been used to correlate field measurements of longshore current velocity and related parameters, including wave period, breaker height and angle, beach slope, and longshore wind velocity. The resulting regression equations demonstrate the relative importance of the independent variables. However, they are generally of use as current prediction equations only at the site where the data were collected.

The most promising analytical approach to longshore current prediction is based on the work of Longuet-Higgins (1970). He equated the longshore component of excess wave momentum flux with the bottom frictional resistance developed by the longshore current. The average longshore current velocity V_l is given by

$$V_l = 2.7 u_m \sin \alpha_b \cos \alpha_b \tag{7.4}$$

where Eq. 7.4 has been calibrated with field data to evaluate friction effects (Komar, 1975). In Eq. 7.4 α_b is the wave breaker angle (see Fig. 7.3) and u_m

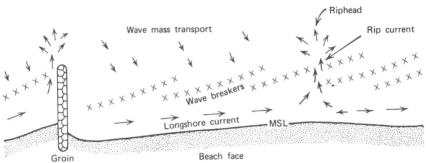

Figure 7.4   Typical wave-generated nearshore circulation.

is the maximum value of the breaking wave orbital velocity, given by

$$u_m = \frac{1}{2} \sqrt{gH_b}$$ (7.5)

where H_b is the breaker height. Although Eq. 7.4 is generally considered to be the best available for longshore current velocity prediction, the difference between predicted and observed velocities can still exceed ± 50 percent.

If a longshore current is intercepted by a headland or a structure (groin, jetty, etc.) oriented normal to the shore, it will be deflected seaward (rip current) and dissipated. A new current will begin to develop downcoast of the obstruction. Even on a long, relatively straight, uninterrupted shoreline, the longshore current will often be interrupted by seaward-flowing rip currents that relieve the continuous longshore accumulation of water in the surf zone. This is demonstrated in Fig. 7.4. The rip current will be situated at a position of lower wave heights and will usually scour a seaward trough, helping to stabilize its location. Shorter period waves tend to produce more frequent but smaller rip currents. See Shepard and Inman (1950) for a further discussion of nearshore circulation cells.

7.4.   LONGSHORE SEDIMENT TRANSPORT

Waves breaking at an angle to the shoreline and the longshore currents generated by these waves combine to cause longshore sediment transport, with most of the transport occurring in the surf zone. The oblique wave runup and gravity generated return flow down the beach face (see Fig. 7.3) cause a "zig-zag" sand particle motion on the beach, with a net motion in the downcoast direction. Sand is also transported in suspension and along the bed (suspended and bed loads, respectively) by the longshore current. Much of the suspended load is placed in suspension by the turbulence generated by wave breaking. Thus the suspended load transport rate is relatively high in the breaker zone. Wave particle orbital motion can also suspend some of the finer sand found seaward of the breaker zone. This sand is then moved downcoast by the longshore current. Bedload transport can occur over the entire zone of the longshore current (see Fig. 7.3) but is also maximum near the breaker zone where turbulence and longshore current velocities are maximum. Finer sand is more likely to be carried as suspended load and coarser sand as bed load.

A laboratory study by Saville (1950) suggests that steeper storm waves cause a predominance of suspended load transport around the breaker zone and lower steepness swell cause bed load transport on the beach to predominate. For both the suspended and bedload transport modes, finer sand sizes will be carried in larger volumes and over longer distances than coarser sizes. One consequence of this is that sediment from a particular point source (e.g. a stream entering the coast) will have a successively finer median diameter at longer distances downcoast from the source. This phenomenon can, at times, be used to establish the predominant drift direction and sediment source for a stretch of coast. When presenting longshore sediment transport rates, it is important to distinguish between the net and gross transport rates at a particular coastal location. The annual directional distribution of wave energy may cause the transport rate in one direction to be so predominant that the gross transport is just slightly greater than the large net transport. On the other hand, annual wave energy may be distributed so that approximately the same volume of sediment is transported in each direction. Then the net transport is near zero but the gross transport may be quite large. The distinction between net and gross transport rates is particularly important, for example, in the design of a sediment bypassing system for a harbor entrance.

Longshore transport rates are usually given as annual volumes of transported sediment, but it must be remembered that the local instantaneous transport rate can be extremely variable, exceeding the average annual transport rate by several times during a storm and falling to zero during periods of low normally incident waves. Also, annual transport rates can be quite variable from year to year owing to fluctuations in wave climate, modifications to coastal structures, and variations in the volume of sediment available from a major source (e.g. a river that has a large flood only every few years and is dry most of the remaining time).

Some average annual net transport rates and directions are listed in Table 7.3 in order to indicate the order of magnitude of longshore transport rates

TABLE 7.3. ESTIMATED NET LONGSHORE TRANSPORT RATES
AND DIRECTIONS

Location	Net Rate (yds^3/yr)	Direction
Sandy Hook, N.J.	477,000	N
Asbury Park, N.J.	200,000	N
Barneget Inlet, N.J.	250,000	S
Ocean City, N.J.	400,000	S
Ocean City, Md.	150,000	S
Hillsboro Inlet, Fla.	75,000	S
Pinellas City, Fla.	50,000	S
Perdido Pass, Ala.	200,000	W
Corpus Christi, Tex.	66,000	S
Santa Barbara, Cal.	280,000	E
Oxnard Plain Shore, Cal.	1,000,000	S
Santa Monica, Cal.	270,000	S
Redondo Beach, Cal.	30,000	S
Camp Pendleton, Cal.	100,000	S
Milwaukee, Wis.	8,000	S
Kenosha, Wis.	15,000	S
Evanston, Ill.	50,000	S

Note: Transport rates obtained from Corps of Engineers project reports
and U.S. Congress House Documents.

on United States beaches. The net rate of 1 million cubic yards per year at
the Oxnard Plain Shore is also essentially the gross rate, as transport is
strongly unidirectional. On the other hand, Corpus Christi is near a nodal
point on the Texas coast where the net transport is near zero. At Corpus
Christi, the annual southerly transport is estimated to be 396,000 cubic yards
per year, while the estimated annual northerly rate is 330,000 cubic yards
per year. Additional transport rates are given in U.S. Army Corps of
Engineers (1973).

The transport rates listed in Table 7.3 were determined primarily by
hydrographic surveys of the volume of sand trapped upcoast or eroded
downcoast of a groin, jetty, or other structure that creates a barrier to
longshore transport. Some of these rates may be underestimates because
most structures do not act as complete barriers to longshore transport.
Transport rate estimates have also been made from dredging records at
harbor entrances that trap sediment, and by special studies using sediment
tracing techniques (see Komar and Inman, 1970).

Functional design of many coastal projects requires estimates of the local net and gross longshore transport rates. These estimates can often be established in one or more of the following ways:

1. If the transport rate has been established at a nearby location with similar shoreline geometry, sand characteristics, and annual wave climate, this rate can be adapted to the project site with possible modification to adjust for local conditions. This requires good engineering judgement as the transport rate and even net direction can change significantly over short distances along the coast.

2. If there are nearby traps to littoral transport and charts showing hydrographic changes over a sufficient period of time, transport estimates can be made. Traps might include man-made structures or geomorphic formations such as headlands and spits. For example, the estimated average annual net longshore transport at Sandy Hook, New Jersey (477,000 cubic yards per year, Table 7.3) is based on hydrographic charts showing the growth of Sandy Hook from 1885–1951.

3. Several field studies have been conducted to relate the longshore sediment transport rate Q_s with the longshore component of wave energy flux (wave power) per unit length of beach at the breaker line P_l. It can be shown (Prob. 2) that

$$P_l = \frac{\rho g}{8} H^2 Cn \cos \alpha_b \sin \alpha_b \qquad (7.6)$$

where H, C, n, and α_b are the values at the breaker line. For a spectrum of waves, $H = H_{rms}$ and C and n are calculated using T_s. If H_s is used in Eq. 7.6, the term on the right must be divided by two because H_s / H_{rms} = 1.41 (see Section 5.2). For sand beaches, a commonly accepted relationship (U.S. Army Coastal Engineering Research Center, 1973), based on several sources of field data, is

$$Q_s = 7.5 \times 10^3 P_l \qquad (7.7)$$

where Q_s is in cubic yards of sand per year and P_l is in foot pounds per second per foot of beach. Relationships like Eq. 7.7 ignore such factors as beach slope, breaker type, and sand characteristics and the plotted data used to develop the relationship show a good deal of scatter. Thus engineering judgement must be used when applying Eq. 7.7.

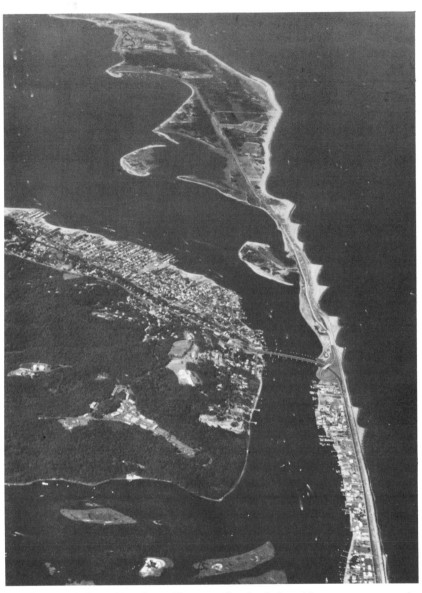

Plate 7.1 Sandy Hook, New Jersey. Transport direction indicated by entrapment at groins. (Courtesy of C. Mason)

7.5. WIND TRANSPORT OF SAND

In addition to the sand transported by waves and littoral currents, large volumes of sand from the beach face and backshore can be transported by the wind. Where there is a large supply of sand from longshore transport, a predominant onshore wind, as is common to many coastal areas, and low coastal topography, wind transported sand can develop a major dune system landward of the winter berm (see Fig. 7.1). There are a variety of coastal dune types (Smith, 1954) and they are found along much of the Atlantic, Gulf, and Pacific coasts of the United States. The dunes of primary interest are those called "foredunes", which are continuous irregular mounds of sand a few meters high situated adjacent and parallel to the beach.

Measurements of foredune growth on the North Carolina, Texas and Oregon coasts (U.S. Army Coastal Engineering Research Center, 1973) indicate onshore wind transport rates of 2–5 cubic yards of sand per foot of beach per year (5–12.5 m^3/m/year). Thus the onshore wind transport rate for a several kilometer long section of coast, can, under appropriate conditions, approach the order of magnitude of typical longshore transport rates. For example, most of the longshore sand transport (approximately 700,000 cubic yards/year) that converges on the nodal point south of Corpus Christi, Texas, is blown inland to form an extensive dune field, with some foredunes having crest elevations above 10 m.

Field and laboratory studies (see Bagnold, 1941; Chepil, 1945) have indicated that three mechanisms are responsible for the transport of sediment by wind:

1. *Saltation.* Particles rise from the surface at a nearly vertical (slightly downwind) angle, travel forward in an arc, and land at a flat (10°–15°) angle at a point 6–10 times the arc height downwind. Upon landing, they may jump or saltate again, or they may dislodge other particles that saltate. The maximum elevation particles achieve is usually less than 1/2 m but may exceed 1 m. The primary cause of the near-vertical particle rise is believed to be hydrodynamic lift. Saltation is usually the predominant mode of sand transport by wind, often accounting for up to 80 percent of the total transport load.

2. *Surface creep.* About 25 percent or less of the wind load is transported by sliding or rolling of particles in continuous contact with the bed. This involves the larger sand grains and the driving forces are wind stress and the impact of saltating grains.

3. *Suspension.* Owing to the low relative density of air, a negligible volume of sand particles having diameters larger than 0.1 mm is carried by

turbulent suspension. Dust and other fine particle sizes not commonly found on a beach can be transported many kilometers and at high altitudes in turbulent suspension.

There is a threshold wind velocity below which sand will not be transported (e.g., approximately 5 m/sec for dry, fine sand). The threshold velocity and subsequent sand transport rate depend on the sand size distribution, moisture content of the sand bed, wind turbulent velocity profile, wind field gustiness, bed slope, and vegetation. When sand is transported in the predominant modes of saltation and surface creep, particle motion is substantially slower than the wind velocity.

Several semi-empirical predictor equations for wind transport rate have been developed (see O'Brien and Rindlaub, 1936; Bagnold, 1941; Kawamura, 1951; and Zingg, 1953). Bagnold's equation is most commonly used. He employed a momentum analysis to evaluate the saltation load and added 25 percent for surface creep. His equation for dry sand is

$$q_s = C\sqrt{\frac{d}{0.25}}\ \rho_a u_*^3$$

where q_s is the rate of sand transport per unit width, d is the median grain diameter in mm, u_* is the shear velocity $\sqrt{\tau_s/\rho_a}$, and C is a coefficient equal to 1.5 for nearly uniform (well-sorted) sand, 1.8 for naturally graded sand, and 2.8 for sand with a wide range of grain diameters. The shear velocity is evaluated from the Karman-Prandtl form of turbulent velocity profile equation or from an equation similar to Eq. 4.23, with the appropriate drag coefficient. Combination of Eqs. 7.8 and 4.23 suggests that the wind transport rate is a function of the wind speed cubed.

A well-developed foredune system can serve two important functions. First, it acts as a seawall to prevent wave and flooding damage landward of the duneline during periods of high water level caused by storms. Second, a coastal dune field is a reservoir of beach sand that can supply sand when the beach is eroded by storm wave attack.

Thus development of a strong foredune system is desirable where an adequate supply of sand is being transported landward by the wind. Growth of dunes can be encouraged and controlled by the installation of semiporous fencing or by the planting of vegetation (particularly beach grasses) to trap sand. Both are particularly effective because of the saltation and surface creep transport mechanisms, which limit sand transport to the region of a meter or less above the ground. Recommended practices for fence construction and grass selection, planting, and care are covered in the Shore

Protection Manual (U. S. Army Coastal Engineering Research Center, 1973).

When planted in sufficient quantity, grasses will continuously trap sand as they grow and the dunes increase in size. Profile data taken at several beach grass planting sites shows dune crest elevations growing at an average rate of about 1/2 m per year for several years to reach elevations of 3–8 m. Fences are less desirable because fencing must be added as dunes grow and the fences deteriorate and become aesthetically less pleasing.

7.6. SEDIMENT TRANSPORT BUDGET

In some areas (e.g. along the southwest coast of Texas) continuous longshore transport of sand can occur over distances as great as 100–200 km. However, in many areas, sand is only transported short distances alongshore from its source or sources before being deposited at one or more semipermanent locations known as sinks. These self-contained coastal regions are often called physiographic units. Whether the physiographic unit is large or small, project planning and design often require development of a sediment transport budget that accounts for all sources and sinks and quantifies rates of sand transport from the sources to the sinks past points of interest.

Sources of beach sand include:

1. *Rivers.* Atlantic and Gulf coast rivers discharge sediment to the coast on a fairly continuous basis, but many Pacific coast rivers and streams are ephemeral and may discharge a large sediment load only during a period of a few days every few years. Much of the sediment load from rivers is finer than the fine sand size and stays in suspension until deposited offshore. If the river enters a coastal embayment, most of the sand size fraction of the sediment load will be deposited before reaching the littoral zone. Dams and erosion control programs can also greatly diminish the value of a river as a source of beach sand.

2. *Beach and cliff erosion.* Often, the main source of sand is an eroding upcoast cliff and/or beach. Beaches supply sand when the wave and longshore current transport capacity at a point exceeds the supply of sand from updrift sources to that point. Beach erosion may be essentially continuous, but it usually occurs at an increased rate during storms when cliff erosion is most common.

3. *Artificial beach nourishment.* In many instances, the most economical way to maintain an eroding beach is by artificial nourishment of the beach using sand from some sources such as offshore deposits, bays, dunefields, and so on. The sand is usually placed on the beach face periodically (e.g. every year or two).

Common sinks for sand transported in the littoral zone are:

1. *Tidal inlets.* Harbor, bay and estuary entrances with tide-generated reversing flow can trap large volumes of the sediment transported alongshore. The flood tide drives sediment through the inlet where it is deposited in quiet water. The ebb tide jet may carry sand far enough offshore to be effectively removed from the littoral zone. Sand may also be trapped adjacent to jetties constructed to stabilize the entrance channel.

2. *Wind-blown sand.* Wind might cause a net seaward transport of sand from the dunes to the littoral zone but at most locations, sand is blown predominantly to the dune field from the beach.

3. *Offshore Deposition.* Some authorities believe offshore sand deposits can be a source of beach sand with the sand being transported onshore by waves. However, in most cases, storm waves will deposit beach sand sufficiently far offshore that all of the sand will not be returned by subsequent low steepness swell, thus causing a net loss to the littoral zone.

4. *Spits, Tombolos, Other Formations.* Depositional features such as spits or hooks (*H*, Fig. 7.5) that grow from shore in the downcoast direction, or tombolos (Figs. 7.7, 7.8) that form behind an offshore obstruction to waves, will trap sand and act as sediment sinks.

5. *Submarine Canyons.* Particularly on the Pacific coast of the United States, a number of submarine canyons cut across the continental shelf with their heads situated very near shore. A portion of the longshore transport is deposited into the canyon head and subsequently transported into deeper water.

6. *Mining.* Sand is a valuable natural resource in many coastal areas. As a consequence it has been mechanically removed from the beach for use elsewhere.

Figure 7.5 is a hypothetical shoreline exposed to waves having the predominant direction shown. Assume that the net sediment transport is almost equal to the gross transport and is in the direction D to H. The longshore transport past headlands A and C will be small, so a pocket beach (B) will form primarily from locally derived sediments, and the shoreline at B will be oriented parallel to the predominant breaker wave crest pattern developed by refraction and diffraction. At D the longshore transport capacity must be satisfied by available sand. The result will be erosion of the beach at a decreasing rate in the downdrift direction until the longshore transport capacity is satisfied. Additional shoreline erosion may occur owing to wind transport and losses offshore during storms. Sand contributed by shoreline erosion plus that discharged by the stream (G) will be transported to form the spit at H. Topographic and hydrographic measurements of shore profile change, and dune and spit growth, combined with estimates of beach

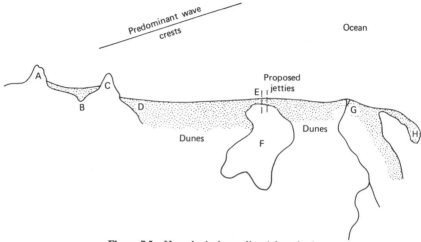

Figure 7.5 Hypothetical coastline (plan view).

size sediment discharge from the stream over a period of time, should balance and provide a sediment transport budget for the physiographic unit between shore points D and H.

An inlet is to be dredged at E and stabilized with jetties to allow controlled tidal flow into and out of bay F. Design of this inlet would require an estimate of the longshore transport rate at E. The annual net transport can be determined from the sediment transport budget. If sufficient wave climate data (e.g. daily average height, period, and direction of waves) are available for a year or more, estimates of both the net and gross longshore transport rates can also be made using Eqs. 7.6 and 7.7. The inlet will function as a sediment sink and modify the sediment transport balance. There will be a local reorientation of the shoreline adjacent to the jetties and additional shoreline erosion downcoast of the inlet. Artificial sediment bypassing may be required at the inlet.

7.7.   COASTAL STRUCTURES

Several types of structures, having a variety of functions, have been constructed alongshore and at harbor and other tidal entrances. These interact with coastal zone processes to redirect longshore and tidal currents and to cause the erosion and/or deposition of sediment.

For convenience, these structures can be classified into three groups: (1) structures built more or less perpendicular to the coast and usually attached to the shore; (2) structures built offshore and approximately parallel to the shore; and (3) structures built on the beach face approximately parallel to the shore.

The first group includes groins for beach stabilization, jetties for harbor and tidal entrance control, and breakwaters to protect a harbor or beach area from wave attack. Jetties protect navigation channels from waves and shoaling by littoral material, they direct channel currents, and they prevent channel migration.

The second class consists primarily of breakwaters (with crowns that are submerged or above water) for shore protection and the development of vessel mooring areas.

A majority of the structures in the last class are seawalls and revetments built at the land-sea interface or up on the beach face to protect shorelines from storm flooding and from beach erosion due to wave attack.

Of concern herein will be the interaction of coastal zone processes with these basic classes of coastal structures. For detailed functional and structural design information, see the U.S. Army Coastal Engineering Research Center (1973) and the various coastal project design reports of the U.S. Army Corps of Engineers.

Figure 7.6a shows the shoreline configuration that will develop adjacent to a structure placed normal to shore and exposed to waves from the direction shown for a sufficient period of time. There will be an upcoast accumulation

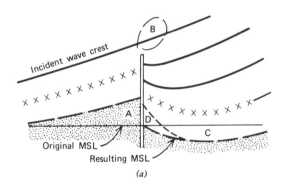

(a)

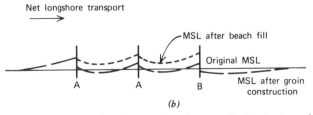

(b)

Figure 7.6 Effect of structures placed normal to shore. a. Single structure; b. system of structures.

of sediment (A) plus the deposition of sediment at point B and further offshore owing to the rip current that will develop along the upcoast face of the structure. Downcoast of the structure (C) the beach will erode at a rate approximately equal to the rate of sediment deposition at A and B. Both upcoast and downcoast of the structure, the shoreline will attempt to adjust so it is parallel to the incoming wave crest positions as affected by refraction and diffraction. Some material will naturally bypass the structure, so the rates of deposition or erosion will be somewhat less than the longshore transport rate. The amount of natural bypassing will depend on the structure length and permeability as well as the beach slope and wave breaking depth, which establish the width of the longshore transport zone and thus the degree of interference to sand transport caused by the structure.

Usually, the wave direction, height, and period are continually changing so equilibrium between the wave crest and shoreline orientation is never completely achieved. The beach is continually adjusting to the changing wave characteristics. However, if the waves come from one predominant direction with only occasional reverses, the resulting shoreline will closely approximate that shown in Fig. 7.6. Waves from the other direction would drive sediment back toward the structure to form a small fillet at D that would be difficult to remove when waves return to the predominant direction. This is particularly true if the structure extends a great distance from shore.

A shoreline response similar to that shown in Fig. 7.6a would also develop adjacent to a pair of parallel jetties constructed at the entrance to a harbor or bay. A portion of the sediment that moves past the offshore end of the upcoast jetty would be transported further offshore by the ebb tidal flow or into the entrance channel by the flood tidal flow.

A common method of preventing beach erosion or rebuilding eroded beaches is to construct a series of structures (groins) normal to shore (Fig. 7.6b) to trap and hold existing longshore transport and/or to be artificially filled with sand. A system of groins can be constructed one section at a time by beginning at the downcoast end and adding new groins as the spaces between the older groins fill with sand. If the entire system is constructed at one time, the updrift groins will fill first; the shoreline between the remaining groins will adjust to the incident waves and then sequentially fill as sediment fills and then passes the upcoast groins. Remember, beach erosion will occur downcoast of the groin system at a rate approximately equal to the rate of deposition in the system. Also, a system of groins will not prevent the loss of sand to offshore during storm wave attack.

The common ratio of groin spacing to length (original MSL shoreline to seaward tip) varies between 1.5 : 1 and 4 : 1. Groins are typically constructed to a point offshore where storm waves break and with a crown elevation

Plate 7.2 Groins at Westhampton, Long Island, New York. (Courtesy of U.S. Army Coastal Engineering Research Center)

equal to that of the winter berm elevation. A design engineer must consider the annual range of incident wave conditions and, from this, anticipate the resulting range of shoreline positions that will develop. It is important that the groins not be flanked by erosion at the landward end, particularly when newly constructed upcoast groins temporarily deny littoral transport to a downcoast segment of the groin system (A, Fig. 7.6b) or when extensive erosion occurs downcoast of the last groin in the system (B, Fig. 7.6b).

The effect of a structure constructed offshore and parallel to shore is demonstrated by the shoreline response (Fig. 7.7) to the breakwater constructed at Santa Monica, California, in 1933-34 (Handin and Ludwick, 1950). The 610-m long breakwater was constructed 610 m seaward of the original shoreline in water 8.5 m deep. Net longshore sediment transport is to the southeast at an average rate of 270,000 cubic yards per year. Waves directly approaching the breakwater are intercepted and adjacent waves are diffracted into the lee of the breakwater. This causes deposition of sand owing to decreased longshore energy flux behind the breakwater and shaping of the deposited sand tombolo by the diffracted wave crests. During the 1933-48 period all beach profile changes occurred landward of the 8-m depth contour, which remained essentially unchanged. Note the upcoast extent of effects on the shoreline. Moving alongshore southward toward the breakwater, the median sand grain diameter is found to be fairly constant until it rapidly increases at a point adjacent to the north tip of the breakwater. Behind the breakwater the median sand grain diameter then rapidly decreases to the lowest values for this coastal segment.

Santa Monica

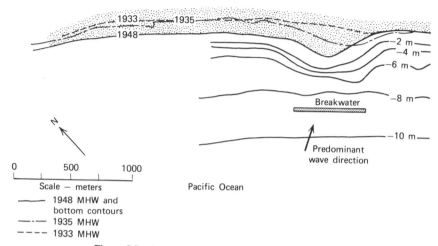

Figure 7.7 Beach changes at Santa Monica, California.

Plate 7.3 Entrance channel to Channel Islands Harbor, California. Note wave diffraction and sand deposition behind breakwater. (Courtesy of U.S. Army Coastal Engineering Research Center)

In some areas a series of short, shore-parallel structures have been constructed to form a segmented offshore breakwater for the purpose of protecting or extending a section of eroding shoreline. The openings in the breakwater allow some additional wave energy into the leeward area to diminish the tendency for a tombolo to form. Openings also increase water circulation.

Some offshore breakwaters have been constructed with their crown near or below MSL. This allows a portion of the incident wave energy to be transmitted over (directly or by wave regeneration by the overtopping water) the breakwater and to the beach. Deposition of some sand will occur in the lee, but the breakwater is not a complete barrier to longshore transport. These structures do create a possible navigation hazard.

Recently, Silvester (Silvester and Ho, 1974) recommended the use of offshore breakwaters to act as headlands and to cause the formation of a series of small, scalloped, crenulate-shaped bays as shown in Fig. 7.8. Again, the resulting shoreline parallels the wave crest orientation due to refraction and diffraction in the lee of the breakwaters. If there is sufficient longshore sediment transport and/or the breakwaters are placed sufficiently close to the original shoreline, the resulting stable scalloped shoreline will develop from natural processes. Otherwise, artificial beach nourishment will be required. Effectively, incident wave energy is spread over a greater length of shoreline with the lowest wave energy approaching shore in the region upcoast and leeward of the breakwaters.

Seawalls and revetments are structures constructed alongshore at the beach face to assist the shoreline in resisting wave attack. Seawalls are usually more massive structures designed to resist larger (ocean) waves and, perhaps, flooding during storm surge. A revetment (e.g. Fig. 6.7) is essentially an armoring of a beach face or bluff to resist wave attack in low to moderate wave climate areas such as interior bays.

Seawalls and revetments only protect the shore landward of the structure and typically have little effect on adjacent upcoast and downcoast areas.

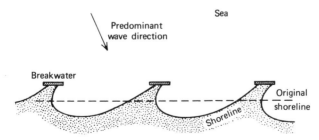

Figure 7.8 Artificial headlands for shore stabilization.

However, if they are used to maintain a section of the shoreline in an advanced position, this outward jutting section of the shore will act as a headland and may trap some portion of the sediment being transported alongshore. Otherwise, sediment is completely free to be transported alongshore in front of the seawall or revetment. Their purpose is to retard beach or bluff erosion during storm wave attack and, where there is insufficient sediment in transport to satisfy the longshore transport capacity of waves.

The upcoast and downcoast ends of a seawall or revetment must tie into a noneroding portion of the shore or must be protected by end walls so the structure is not flanked by the erosion of adjacent beaches.

The beach seaward of the structure will continue to erode toward some equilibrium profile that depends on the incident wave climate, structure geometry and supply of longshore sand transport. The scour depth at the toe of the structure is accentuated by wave reflection from the structure which is greater, for example, for a smooth, impervious, vertical-faced, concrete seawall than for an inclined, rubble-mound revetment. To prevent undermining of the structure, the toe must be placed sufficiently deep or be protected by a cutoff wall (see Fig. 6.7).

7.8. BEACH NOURISHMENT AND SEDIMENT BYPASSING

Many shore stabilization and beach improvement problems can be solved most economically by a periodic artificial transfer of sand. These efforts usually fall into two categories: the nourishment of beaches with sand from a nearby onshore or offshore borrow area, and the transport of sand to bypass obstructions to longshore transport (e.g. coastal harbors and jettied tidal entrances to bays and interior harbors). Sand transfer is accomplished either hydraulically by some form of mobile dredging plant and discharge line or, less commonly, by mechanical means such as a clamshell or dragline feeding trucks or barges. Fundamentals of artificial beach nourishment and sand bypassing will be presented; specific techniques and examples are given by the U.S. Army Coastal Engineering Research Center (1973).

Erosion of a section of shoreline (beach or bluff) can be prevented by artificial nourishment of the shore with sand, to develop a protective beach. If an adequate sand source can be found this form of shore protection can be less expensive than purely structural solutions. Beach nourishment is also likely to benefit rather than harm adjacent downcoast beaches, because sand transported from the nourished area feeds these beaches. A single beach nourishment rarely provides a permanent solution to a beach erosion problem, so periodic renourishment is usually necessary. Often, a combination of beach nourishment and construction of groins or offshore breakwaters provides the most economical means of protecting a section of the shore. Artificial nourishment can also be used to develop recreational beaches in areas with insufficient natural supplies of sand.

A cost-effective source of borrow material for beach nourishment must have a suitable particle size distribution for the wave climate and beach slope at the nourishment site. It must also be close to the site to keep sand transport costs down, be suitable to the site environmental and recreational requirements (e.g. clean, not too coarse), and its removal must not cause ecological and other problems at the borrow area. Some common borrow areas are onshore quarries, bays and lagoons, dune fields, and offshore sites including glacial deposits, former drowned beaches, and deposits seaward of tidal inlets.

The volume of sand needed for the initial nourishment of a section of beach depends on (1) the desired beach profile, which depends on the proposed use of the beach (e.g. recreation, shore protection); (2) the overfill needed to allow for subsequent beach erosion; and (3) the overfill needed to allow for the initial rapid removal of the finer portion of the fill material by wave action.

Allowance for beach erosion depends on the rate of beach recession and the frequency and volume of periodic beach renourishment planned. The beach recession rate can be estimated from the past behavior of the beach but may be more rapid for the nourished beach due to its extended geometry and particularly if the fill material is finer than the native beach material. Recession of the nourished beach can be reduced, of course, by the construction of groins or offshore breakwaters to stabilize the beach.

A critical factor in planning a beach nourishment project is a comparison of the native beach sand size distribution (based on a composite surface sample) with the size distribution of the borrow material (ascertained from core samples). Since the native sand is eroding or, at best, marginally stable, and if no protective structures are to be used, it is desirable to use fill material with the same or larger median diameter than the native sand. When estimating the necessary overfill volume, it is usually assumed that the portion of the fill size distribution finer than the particle sizes in the native size distribution will be immediately lost from the nourished beach. James (1975) surveys the various existing analytical techniques for determining the overfill volume needed, based on a comparison of the natural and borrow material's composite particle size distributions.

Commonly, beach fill is placed directly on the beach face and distributed by earth moving equipment. If there is a strong unidirectional longshore transport it may be economically and functionally preferable to stockpile the fill in the surf zone at the updrift end of the beach to be nourished. If a stockpile of sand is to be placed on the eroding shore downdrift of a tidal inlet or harbor entrance, care must be taken in locating the stockpile so that during the periods of transport reversal excessive amounts of sand are not transported back into the entrance.

A variation on the nourishment of an eroding beach with sand borrowed from a distant source is the mechanical transfer of sand from an updrift region of deposition, past an obstruction such as a harbor or jettied inlet, to a downdrift location where the sand can continue to move as part of the natural longshore transport regime. Mechanical bypassing is usually accomplished on an intermittent basis with a floating dredge and a discharge pipeline, but has also been done on a continuous basis with a stationary pumping plant and associated intake and discharge lines. There will be some natural bypassing of most obstructions so it is usually only necessary to mechanically bypass less than 100 percent of the longshore transport. It is important to locate the sand collection and release points so losses owing to deposition inside or offshore of a tidal entrance are eliminated to the greatest extent possible. When the gross transport rate greatly exceeds the net rate it is desirable to design the bypassing system so only the net transport rate has to be satisfied. This would involve some arrangement where sand trapped at an obstruction can be naturally transported back upcoast during periods when the wave direction is reversed.

Other considerations in the design of a sand bypassing system are:

1. Longshore transport rates are very irregular. Peak daily rates will be several times the annual average daily rate and these peak rates will usually occur during periods of higher (storm) wave activity when the bypassing operation may be shut down. A drift deposition basin that is large enough to handle peak transport rates and still be completely accessible to the bypassing mechanism is needed when bypassing is continuous.

2. Most floating dredges are sensitive to wave attack so the deposition basin should be protected from waves if the dredge is to be used economically. Otherwise, the dredge can be used only during periods of low waves.

3. The width of the longshore transport zone must be determined so all material can be trapped and the dimensions of the deposition zone can be estimated. This width depends on the beach slope, wave climate and tidal range.

Figure 7.9 illustrates the more common types of sand bypassing systems in use. The oldest approach (Fig. 7.9a) used in the United States was to accept

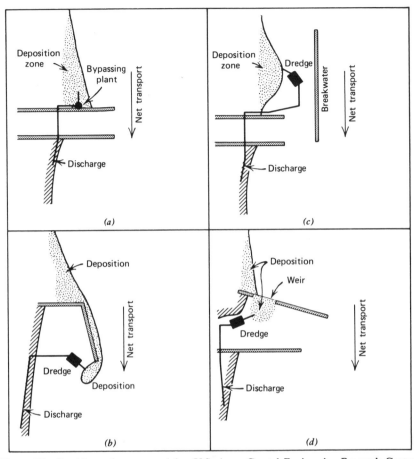

Figure 7.9 Sand bypassing systems (after U.S. Army Coastal Engineering Research Center, 1973).

the natural deposition zone updrift of a jetty and transfer the sand from the deposition zone by clamshell and truck, or by a fixed dredging plant installed on the updrift jetty. This plant consists of a boom supporting an intake pipe that can be swung in an arc over the deposition zone. The intake pipe is connected to a centrifugal pump and discharge line that crosses the inlet.

At a harbor with a shore connected breakwater (Fig. 7.9b), the drift eventually moves into the harbor entrance and develops a spit in the deposition zone. The breakwater and spit will protect a dredge working in their lee to maintain the harbor mooring area and entrance channel.

Figure 7.9c shows a very effective but expensive bypassing system in which an offshore breakwater develops a deposition zone and sheltered area for a dredge. The breakwater can also be placed to provide additional shelter to the entrance navigation channel.

The system depicted in Fig. 7.9d consists of a weir (crest elevation at or near MSL) at the shoreward end of the upcoast jetty, which is oriented to allow for a protected deposition basin in the lee of the weir and jetty. The low crest elevation allows sand to move over the weir and into the deposition basin. Examples of the four systems in Fig. 7.9 are South Lake Worth Inlet, Florida (Watts, 1953); Santa Barbara Harbor, California (Wiegel, 1959); Channel Islands Harbor, California (Herron and Harris, 1966); and Mason-boro Inlet, North Carolina (Magnuson, 1966), respectively.

7.9. SUMMARY

This chapter has covered elements of coastal sediment and beach character-istics; coastal zone processes, including the transport of sand by waves, currents, and wind; the interaction of these transport processes with coastal structures; and the major structural and nonstructural means of solving shoreline erosion and sediment deposition problems. Control of shore erosion and prevention of sedimentation at harbors and other coastal areas con-stitutes the major class of problems facing coastal engineers in the United States and many other coastal countries.

7.10. REFERENCES

Bagnold, R. A. (1941), *The Physics of Blown Sand and Desert Dunes*, Methuen, London, 265 p.

Bascom, W. N. (1951), "The Relationship Between Sand Size and Beach-Face Slope," *Transactions, American Geophysical Union*, December, pp. 866–874.

Chepil, W. S. (1945), "Dynamics of Wind Erosion," *Soil Science*, Vol. 60, pp. 305–320, 397–411, 475–480.

Colby, B. C. and R. P. Christensen (1956), "Visual Accumulation Tube for Size Analysis of Sands," *Journal, Hydraulics Division*, American Society of Civil Engineers, June, paper 1004.

Dean, R. G. (1973), "Heuristic Models of Sand Transport in the Surf Zone," *Conference on Engineering Dynamics in the Surf Zone*, Sydney, pp. 208–214.

Galvin, C. J. (1967), "Longshore Current Velocity: A Review of Theory and Data," *Reviews of Geophysics*, Vol. 5, pp 287–304.

Griffiths, J. C. (1967), *Scientific Method in Analysis of Sediments*, McGraw-Hill, New York, 508 p.

Handin, J. W. and J. C. Ludwick (1950), "Accretion of Beach Sand Behind a Detached Breakwater," Tech. Memo. 16, U.S. Army Beach Erosion Board, Washington, D. C., 13 p.

Herron, W. J. and R. L. Harris (1966), "Littoral Bypassing and Beach Restoration in the Vicinity of Port Hueneme, California," *Proceedings, Tenth Conference on Coastal Engineering*, Tokyo, pp. 651–675.

Ingle, J. C. (1966), *The Movement of Beach Sand*, Elsevier, New York, 221 p.

Inman, D. L. (1952), "Measures for Describing the Size Distribution of Sediments," *Journal, Sedimentary Petrology*, Sept., pp. 125–145.

James, W. R. (1975), "Techniques in Evaluating Suitability of Borrow Material for Beach Nourishment," Tech. Memo. 60, U.S. Army Coastal Engineering Research Center, Ft. Belvoir, Virginia, 81 p.

Kamel, A. and J. W. Johnson (1963), "Tracing Coastal Sediment Movement by Naturally Radioactive Minerals," *Proceedings, Eighth Conference on Coastal Engineering*, Mexico City, pp. 324–330.

Kawamura, R. (1951), "Study on Sand Movement by Wind," Univ. of Tokyo, Institute of Science and Technology Report, Vol. 5, pp. 94–112.

Komar, P. D. (1975), "Nearshore Currents: Generation by Obliquely Incident Waves and Longshore Variations in Breaker Height," in *Nearshore Sediment Dynamics and Sedimentation* (Hails and Carr, Editors), Wiley, New York, pp. 17–45.

Komar, P. D. and D. L. Inman (1970), "Longshore Sand Transport on Beaches," *Journal, Geophysical Research*, Vol. 75, pp. 5914–5927.

Krumbein, W. C. (1936), "Application of Logarithmic Moments to Size Frequency Distribution of Sediments," *Journal, Sedimentary Petrology*, pp. 35–47.

Krumbein, W. C. (1954), "Statistical Significance of Beach Sampling Methods," Tech. Memo 50, U.S. Army Beach Erosion Board, Washington, D. C., 33 p.

Krumbein, W. C. and G. D. Monk (1942), "Permeability as a Function of the Size Parameters of Unconsolidated Sand," Tech. Publ. 1492, *Petroleum Technology*, July, pp. 1–11.

Krumbein, W. C. and F. J. Pettijohn (1938), *Manual of Sedimentary Petrography*, Appleton-Century, New York, 549 p.

Krumbein, W. C. and H. A. Slack (1956), "Relative Efficiency of Beach Sampling Methods," Tech. Memo. 90, U.S. Army Beach Erosion Board, Washington, D. C., 52 p.

Longuet-Higgins, M. S. (1970), "Longshore Currents Generated by Obliquely Incident Sea Waves," *Journal, Geophysical Research*, Vol. 75, pp. 6778–6789, 6790–6801.

Magnuson, N. C. (1966), "Planning and Design of a Low-Weir Section Jetty at Masonboro Inlet, North Carolina," *Proceedings, Santa Barbara Coastal Engineering Specialty Conference*, pp. 807–820.

Mahlig, W. C. and A. E. Reed (1972), *Manual on Test Sieving Methods*, American Society for Testing and Materials, Philadelphia, 43 p.

O'Brien, M. P. and B. D. Rindlaub (1936), "The Transportation of Sand by Wind," *Civil Engineering*, Vol. 6, pp. 325–327.

Saville, T. (1950), "Model Study of Sand Transport Along an Infinitely Long Straight Beach," *Transactions, American Geophysical Union*, Vol. 31, pp. 555–565.

Schlee, J. (1966), "A Modified Woods Hole Rapid Sediment Analyser," *Journal, Sedimentary Petrology*, June, pp. 403–413.

Seelig, W. N. and R. M. Sorensen (1973), "Texas Shoreline Changes," *Journal, American Shore and Beach Preservation Association*, October, pp. 23–25.

Shepard, F. P. and D. L. Inman (1950), "Nearshore Circulation Related to Bottom Topography and Wave Refraction," *Transactions, American Geophysical Union*, Vol. 31, pp. 196–212.

Silvester, R. and S. Ho (1974), "New Approach to Coastal Defense," *Civil Engineering*, September, pp. 66–69.

Smith, H. T. U. (1954), "Coast Dunes," *Proceedings, Coastal Geography Conference*, Office of Naval Research, pp. 51–56.

Sonu, C. J., J. M. McLoy and D. S. McArthur (1967), "Longshore Currents and Nearshore Topography," *Proceedings, Tenth Conference on Coastal Engineering*, Tokyo, pp. 524–549.

Szuwalski, A. (1970), "Littoral Environment Observation Program in California—Preliminary Report," Misc. Publ. 2–70, U.S. Army Coastal Engineering Research Center, Washington, D. C.

Trask, P. D. (1952), "Source of Beach Sand at Santa Barbara, California, as Indicated by Mineral Grain Studies," Tech. Memo. 28, U.S. Army Beach Erosion Board, Washington, D. C., 24 p.

U.S. Army Coastal Engineering Research Center (1973), *Shore Protection Manual*, 3 Vols., U.S. Government Printing Office, Washington, D. C.

Watts, G. M. (1953), "A Study of Sand Movement at South Lake Worth Inlet, Florida," Tech. Memo 42, U.S. Army Beach Erosion Board, Washington, D. C., 24 p.

Wentworth, C. K. (1922), "A Scale of Grade and Class Terms for Clastic Sediments," *Journal, Geology*, p. 377–392.

Wiegel, R. L. (1959), "Sand By-Passing at Santa Barbara, California," *Journal, Waterways and Harbors Division*, American Society of Civil Engineers, June, pp. 1–30.

Zingg, A. W. (1953), "Wind-tunnel Studies of the Movement of Sedimentary Material," *Proceedings, Fifth Hydraulics Conference*, University of Iowa, pp. 111–135.

7.11. PROBLEMS

1. A sieve analysis of a sand sample showed the following:

Opening Size (mm)	Weight Retained (grams)	% Passing
2.000	0	100
1.414	0	100
1.000	0.3	99.7
0.707	1.7	98.3
0.500	6.2	93.8
0.353	27.8	72.2
0.250	24.1	76
0.177	17.7	
0.125	15.3	
0.088	5.0	
0.062	1.9	

Plot the sample cumulative frequency distribution on log-normal graph paper and determine the phi median diameter, phi mean diameter, phi deviation measure, and phi skewness measure. Estimate the sample permeability.

2. Demonstrate that the longshore component of wave energy transfer (wave power) per unit length of beach at the breaker line is given by Eq. 7.6.

3. A beach consists of quartz sand particles having a median settling diameter of 0.24 mm. Plot, on a graph of wave height H_0 versus period T, the line (band) separating eroding and accreting beach profiles. Use a range of H_0 and T values common to the ocean environment. Discuss the practical significance of the results indicated by this plot.

4. A beach is composed of sand having a median diameter of 0.5 mm. The estimated bottom velocity required to initiate noticeable motion of this sand is 0.25 m/sec. For a wave ($T=6$ sec, $H_0=1.5$ m) shoaling on this beach, at what depth does sand particle motion commence?

5. A wave train with an average period of 8 sec and breaker height of 2 m is approaching the shore and breaking at an angle of $12°$ with the shoreline. What average longshore current velocity is generated? If this wave train represents the average conditions for the day, what volume of longshore sand transport is produced in 24 hr?

6. At a tidal inlet on a N-S oriented coastline, the average annual longshore transport is 300,000 m^3 to the south during the winter and 130,000 m^3 to the north during the summer. The average nearshore beach slope is $1:50$. A 400-m wide navigation channel is to be dredged to a depth of -4 m and protected over its entire length by jetties. Suggest the layout for jetties at this inlet. Discuss any bypassing system included with the jetties.

7. From a study of the details of the coastal hydrographic chart provided by your instructor, discuss the features observed and active coastal zone processes indicated. Include any sources or sinks, sediment transport directions, effects of structures, and so on.

Eight

DIFFUSION IN COASTAL WATERS: SUBMARINE OUTFALL DESIGN

Disposal of municipal and industrial sewage and other liquid wastes through submarine outfall lines to the ocean is done to some extent at most coastal population centers. Varying types and amounts of treatment are given to the effluent waste prior to discharge. The submarine outfall must be located and designed to allow sufficient dilution of the effluent, die-away of bacteria, and settlement of solids, or ecological problems will be created in the receiving water and at adjacent shores. Standards for receiving water quality, though difficult to develop, have been established for most coastal areas. The level of waste treatment and the submarine outfall design are interrelated and based (within limits set by local receiving water standards) on an economic trade-off between waste disposal costs and the decreasing value of receiving waters owing to pollution.

The main marine hydromechanics problem is to predict the dilution that turbulent diffusion will produce for a given effluent pollutant concentration and discharge rate, submarine outfall geometry and location, and receiving water physical and dynamic characteristics. Thus the primary aim of this chapter is to present the fundamentals of turbulent diffusion, with particular reference to submarine outfall design.

8.1. PHYSICAL PROBLEM

Figure 8.1 depicts a submarine outfall discharging an effluent that is less dense than the ambient sea water. Also shown are the major phenomena

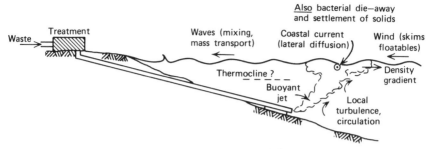

Figure 8.1 Submarine outfall.

that affect the characteristics of the effluent (liquid, solids, organisms) after it is discharged from the outfall. The effluent may be discharged through a simple opening at the end of the pipeline or, to achieve greater initial mixing with receiving waters, it may be discharged through a mainfold consisting of several openings (ports) spaced along the outer portion of the outfall line.

Most sewage has essentially the same density as fresh water, and heated sea water used as cooling water will be less dense than sea water at normal temperatures, so the effluent will usually be buoyant and will rise toward the surface as a turbulent jet. This turbulent jet will entrain receiving water as it rises causing significant dilution of the effluent. The vertical density profile in the ambient receiving water and the possible existence of a thermocline will control the buoyant force on the jet, the resulting rate at which it rises to the surface, and the possibility of the jet attaining buoyant equilibrium before reaching the surface. The effluent jet will generate local turbulence in the receiving water that, along with the existing turbulence, cause mixing and dilution as the jet rises toward the surface.

If upon reaching the surface the effluent is still lighter than the receiving water, there will be a lateral spreading of the effluent owing to the horizontal density gradient. Wind will skim floating solids and transport them (with any odors) in the downwind direction. This may cause severe design constraints if strong onshore winds frequently occur. Wave action will increase mixing (and aeration) near the surface and wave mass transport will cause a slow drift of the effluent in the direction of wave motion.

If there is no coastal current, the effluent will accumulate near the outfall site with only slight spreading and dilution caused by wind, waves, diffusion owing to local turbulence, density differences, and local circulation cells generated by the buoyant jet. One must usually rely primarily on coastal currents for significant dilution of the effluent after it reaches the surface. Lateral turbulent diffusion will cause continuous mixing of the effluent and receiving water as the effluent is transported from the outfall site (and, hopefully, away from shore) by coastal currents.

As time elapses, nonfloating solids, held in turbulent suspension or formed by coagulation (caused by the colder receiving water), will settle out and harmful organisms will die away. Thus it is desirable to increase the effluent travel time between the outfall site and areas where the effluent can cause problems, in order to allow for maximum settling of solids, die-away of organisms, and dilution by turbulent mixing.

As a marine hydromechanics problem, the dilution of a liquid effluent in the ocean will be divided into two separate phenomena. First, the dynamics and resulting dilution of a single turbulent buoyant jet rising from an outfall port will be presented. This will be followed by a derivation and application of the mass balance equation for turbulent mixing in a horizontal current. By combining the effects of these two phenomena, one can predict the resulting dilution of a liquid waste discharged from a submarine outfall.

8.2. TURBULENT BUOYANT JETS

A turbulent effluent jet discharging horizontally into denser homogeneous receiving water and rising to the surface is shown in Fig. 8.2. Y is the vertical distance from the port centerline to the free surface, D is the initial diameter of the circular jet (also the port diameter if no jet contraction), V and ρ are the initial jet velocity and density, ρ_0 is the receiving water density, and S_0 is the surface centerline dilution (volume of effluent and receiving water mixture divided by the volume of effluent in the mixture, i.e. reciprocal of the effluent concentration). The effluent and receiving water can be assumed to have the same kinematic viscosity ν. We also can define an apparent acceleration of gravity g', which is equal to the buoyant force per unit mass of effluent, and is given by (see Prob. 1)

$$g' = \frac{\rho_0 - \rho}{\rho} g \qquad (8.1)$$

Approximate relative dimensions of the plume based on experiments by Rawn and Palmer (1930) and Hart (1961) are shown in Fig. 8.2. Because it is desirable to achieve as much dilution as possible before the effluent reaches the surface, horizontal discharge (rather than vertical or at some angle between horizontal and vertical) is preferable and most common.

The geometry of the plume and the resulting effluent dilution will depend primarily on the ratio of water depth to discharge port diameter, the initial momentum of the effluent jet, and the buoyant force owing to the effluent-

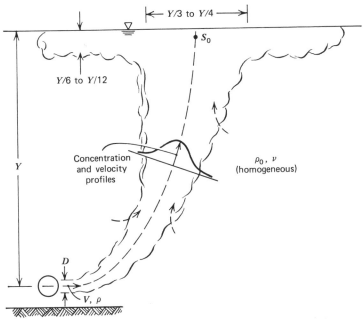

Figure 8.2   Turbulent buoyant jet discharging into homogeneous receiving water.

receiving water density difference. As the effluent rises, receiving water is entrained, the effluent concentration and thus density and buoyant force decrease, and the plume spreads laterally. The effluent concentration and jet velocity are maximum along the plume centerline axis and decrease with distance from the centerline, according approximately to a normal or Gaussian distribution. The lateral velocity gradient enhances the turbulence, which, in turn, acts on the lateral concentration gradient to cause spreading and diffusion of the effluent in the plume.

Further insight into the mechanics of a turbulent buoyant jet can be gained by considering the balance between the buoyant-driving and turbulent-shear resisting force vectors and the resulting change in momentum flux (density × discharge × velocity) of the jet. The buoyant force diminishes with distance along the length of the jet as mixing occurs and as lighter ambient water is encountered closer to the surface. While the buoyant force always acts vertically upward, the turbulent shear force acts opposite to the local flow direction of the curving jet. It also diminishes along the length of the jet as the velocity diminishes.

In the horizontal direction, the momentum flux is maximum at the discharge port. Thereafter, it continuously decreases toward zero owing to opposition by the horizontal component of turbulent shear resistance. Thus the horizontal component of the jet velocity decreases from a maximum at the discharge port toward zero as the jet approaches the surface. Both the density and discharge increase along the jet axis, owing to entrainment of receiving water, so a decrease in momentum flux requires a decrease in jet velocity.

In the vertical direction, initially only the buoyant force acts and causes the jet to gain vertical momentum flux, and thus an upward acceleration component. Along the jet axis, the decreasing buoyant force and increasing vertical shear component, owing to the change in jet direction and increased jet velocity, cause the vertical acceleration to continuously diminish. If the plume is sufficiently long, it is possible for the buoyant force to eventually become less than the turbulent shear force and cause the jet to decelerate and possibly come to rest at some point below the surface. The buoyant force can even become negative if the effluent sufficiently mixes with dense bottom water so its density becomes greater than less-dense surface waters. Thus the plume is likely to be trapped below the surface if a strong thermocline exists to act as a lid. Vertical momentum flux may carry the jet above the point of static density equilibrium, but when this momentum is dissipated, the effluent settles back to the level having the same ambient density, and spreads at this level. Hart (1961) performed laboratory experiments with a buoyant jet discharging into a density stratified receiving water that dramatically demonstrate the various types of behavior of a buoyant jet.

The mechanics of a turbulent buoyant jet have been investigated analytically by Abraham (1963, 1965) and Fan and Brooks (1969), who present solutions for circular and slotted jets discharging at various angles from horizontal to vertical into homogeneous and stratified oceans. Numerical solutions are presented and the results compared with experimental data. Herein, we will only consider the problem of a circular horizontal buoyant jet discharging into homogeneous receiving water. The problem will be approached using dimensional analysis as was done by Rawn, Bowerman, and Brooks (1960).

From the previous discussion, one can reason that

$$S_0 = f(Y, D, g', V)$$

and dimensional analysis yields

$$S_0 = f\left(F, \frac{Y}{D}\right) \tag{8.2}$$

where

$$F = \frac{V}{\sqrt{g'D}} \quad \text{(Froude number)} \qquad (8.3)$$

If the discharge velocity and port diameter are sufficiently large (so the Reynolds number exceeds about 5000) the jet will be fully turbulent, eddy viscosity will predominate over molecular viscosity, which then need not be considered, and flow patterns will be dynamically similar if geometric ratios and Froude numbers are equal (Eq. 8.2). Prototype Reynolds numbers are of the order of 10^5–10^6, and if Reynolds numbers are sufficiently high in laboratory experiments, the relationship given by Eq. 8.2 can be used to present the experimental results.

Figure 8.3, a plot of Y/D versus F for various values of S_0, is from Abraham (1963), and considers experimental data from Rawn and Palmer (1930) and Cederwall (1963). As would be expected, for a given discharge and port diameter (constant F), an increase in water depth (i.e. Y/D) increases the jet centerline dilution at the surface. For a given outfall geometry (Y/D constant) the dilution decreases as the discharge (i.e. F) increases to a certain point, then the dilution increases again with further increases in discharge. Sharp (1968) explains the existance of this minimum

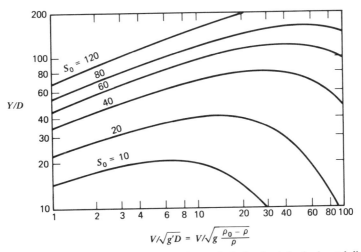

Figure 8.3 Surface dilution versus Froude number and relative depth for horizontal discharge (Abraham, 1963).

dilution point (curve peak) by introducing the concept of a starved jet, in which buoyant forces can generate a greater upward mass transport than provided by the given discharge. This causes a breakdown in plume continuity, gusting, and increased dilution as the discharge F decreases below the minimum dilution point. Above this point, dilution also increases with increasing discharge because the plume length increases owing to increasing initial jet momentum. Abraham (1963, 1965) presents a figure showing the path of the plume axis and dilution at points along the axis as a function of Froude number.

8.3. LATERAL DIFFUSION IN AN OCEAN CURRENT

In order to evaluate the characteristics of an effluent at some point downcurrent of the discharge site, one must determine the travel time to that point as well as the rate of lateral diffusion that occurs during travel. The product of the time rates of bacterial die-away and mixing by lateral diffusion, and the travel time yield the final bacterial concentration at the point of interest.

As a rising effluent plume (or plumes from a multi-port diffuser) reaches the surface and is transported away by a current, a relatively homogeneous effluent field rapidly develops, having an initial average dilution S_a. The effluent field will have a width w that is approximately equal to the length of the diffuser section (plus $Y/3$ to $Y/4$) projected along a line normal to the current direction. The field vertical thickness h is difficult to predict but can be estimated from the values given in Fig. 8.2. The field thickness may be limited to some smaller value, owing to receiving water density stratification, so available experience with local conditions should be relied upon in estimating the field thickness. From continuity of total flow (effluent plus receiving water)

$$S_a Q = Uwh \qquad (8.4)$$

where U is the current velocity and Q is the rate of effluent discharge from the outfall.

It is unlikely that the surface centerline dilution S_0 in the plume (or $2S_0$ to approximate the average dilution across the plume or plumes) will equal the initial current dilution S_a. If $S_a > 2S_0$, excess diluting water is available from the ocean current. Initially, there will be local areas with a dilution less than S_a but, as the effluent is carried away, turbulence will cause the dilution across most of the field to approach S_a. If $S_a < 2S_0$, all effluent reaching the surface will not be removed by the current, so effluent will accumulate near the outfall. If this occurs for design conditions, the outfall design should be modified (see Prob. 2 and 3) to eliminate this accumulation of effluent.

As an effluent plume is carried from the outfall site by ocean currents, dilution continues, owing primarily to lateral (i.e. normal to plume axis) turbulent diffusion. An equation that defines the turbulent diffusion in an ocean current can be derived by writing the mass balance equation for the diffusing substance (effluent) as it is carried by the surrounding flow through a fixed differential control volume, as shown in Fig. 8.4. In Fig. 8.4, c is the instantaneous concentration of diffusing substance (i.e. mass of effluent per total mass of effluent plus receiving water), which is the reciprocal of the dilution as previously defined.

The inflow minus outflow of diffusing effluent must equal the time rate of increase of effluent in the control volume. Writing this balance for flow in the three coordinate directions and the resulting accumulation of mass in the control volume, and then dividing through by the elemental volume yields

$$\frac{\partial (\rho c)}{\partial t} + \frac{\partial (\rho c u)}{\partial x} + \frac{\partial (\rho c v)}{\partial y} + \frac{\partial (\rho c w)}{\partial z} = 0$$

As the change in effluent density owing to diffusion is small, assume density is constant. Thus

$$\frac{\partial c}{\partial t} + \frac{\partial (c u)}{\partial x} + \frac{\partial (c v)}{\partial y} + \frac{\partial (c w)}{\partial z} = 0 \qquad (8.5)$$

The effects of molecular diffusion are small and can be neglected. To define the turbulent motion let the instantaneous concentration and velocity at a point be the sum of time-average $(\bar{c}, \bar{u}, \bar{v}, \bar{w})$ and fluctuating (c', u', v', w')

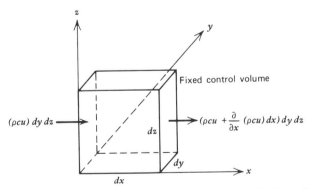

Figure 8.4 Fixed control volume (only x-components of entering and exiting effluent flux are shown).

components to yield $c = \bar{c} + c'$, $u = \bar{u} + u'$, and so on. These terms are then substituted into Eq. 8.5 and the individual terms are expanded. For example,

$$\frac{\partial(cu)}{\partial x} = \frac{\partial(\bar{c}\bar{u})}{\partial x} + \frac{\partial(c'u')}{\partial x} + \frac{\partial(\bar{c}u')}{\partial x} + \frac{\partial(c'\bar{u})}{\partial x} \qquad (8.6)$$

The time average of the expanded form of Eq. 8.5 reduces to

$$\frac{\partial c}{\partial t} + \frac{\partial(\bar{u}\bar{c})}{\partial x} + \frac{\partial(u'c')}{\partial x} + \frac{\partial(\bar{v}\bar{c})}{\partial y} + \frac{\partial(v'c')}{\partial y} + \frac{\partial(\bar{w}\bar{c})}{\partial z} + \frac{\partial(w'c')}{\partial z} = 0 \quad (8.7)$$

because terms like the third and fourth terms on the right side in Eq. 8.6 become zero when averaged over a short period of time. This is because u' or c' are zero by definition when a time average is taken. However, the product of u' times c' will typically not equal zero when time averaged.

Expanding the second, fourth, and sixth terms in Eq. 8.7, and remembering that from continuity of total mass transport,

$$\frac{\partial \bar{u}}{\partial x} + \frac{\partial \bar{v}}{\partial y} + \frac{\partial \bar{w}}{\partial z} = 0$$

Eq. 8.7 becomes

$$\frac{\partial \bar{c}}{\partial t} + \bar{u}\frac{\partial \bar{c}}{\partial x} + \bar{v}\frac{\partial \bar{c}}{\partial y} + \bar{w}\frac{\partial \bar{c}}{\partial z} + \frac{\partial(u'c')}{\partial x} + \frac{\partial(v'c')}{\partial y} + \frac{\partial(w'c')}{\partial z} = 0 \quad (8.8)$$

The first term on the left represents the time rate of change of average effluent concentration at a point, the next three terms represent the convective effluent mass transport owing to concentration gradients in the component flow directions, and the last three terms represent the diffusion of mass in the three component directions owing to turbulent mixing.

It is convenient to replace the turbulent fluctuation terms in Eq. 8.8 by the product of a diffusion coefficient and the concentration gradient, that is,

$$u'c' = -D_x\frac{\partial \bar{c}}{\partial x}$$

where the minus sign is included because a negative concentration gradient in a turbulent field produces a positive mass transport. The diffusion coefficient (or eddy diffusivity) is variable in space and time and depends on the turbulence level. It is closely related to the coefficient of eddy viscosity.

Thus Eq. 8.8 becomes

$$\underbrace{\frac{\partial \bar{c}}{\partial t}}_{(1)} + \underbrace{\bar{u}\frac{\partial \bar{c}}{\partial x}}_{(2)} + \underbrace{\bar{v}\frac{\partial \bar{c}}{\partial y}}_{(3)} + \underbrace{\bar{w}\frac{\partial \bar{c}}{\partial z}}_{(4)} - \underbrace{\frac{\partial}{\partial x}\left(D_x \frac{\partial \bar{c}}{\partial x}\right)}_{(5)} - \underbrace{\frac{\partial}{\partial y}\left(D_y \frac{\partial \bar{c}}{\partial y}\right)}_{(6)} - \underbrace{\frac{\partial}{\partial z}\left(D_z \frac{\partial \bar{c}}{\partial z}\right)}_{(7)} = 0 \quad (8.9)$$

which defines the change in concentration at a point owing to mixing in three-dimensional turbulent flow.

In order to apply Eq. 8.9 we must make the following basic assumptions (let z represent the vertical axis):

1. Flow is steady, so term 1 reduces to zero.
2. Owing to limitations imposed by the bottom, air-sea interface, and possible vertical density stratifications, vertical mixing is negligible, and terms 4 and 7 reduce to zero.
3. Flow is in the x-direction ($\bar{v}=0$), so term 3 reduces to zero.
4. Turbulent mixing in the flow direction is negligible, so term 5 reduces to zero. Owing to convection of mass in the flow direction, the concentration gradient is much lower in that direction than it is normal to flow, and term 5 ≪ term 6.
5. The coefficient of eddy diffusion D_y is essentially constant for a particular flow situation (i.e. outfall design).

Thus Eq. 8.9 becomes

$$\bar{u}\frac{\partial \bar{c}}{\partial x} - D_y \frac{\partial^2 c}{\partial y^2} = 0 \qquad (8.10)$$

Equation 8.10 thus equates the decrease in effluent concentration in the direction of flow to the lateral diffusion of effluent by turbulent mixing. As other components of dilution (assumptions 2 and 4) are neglected, Eq. 8.10 should predict conservative (i.e. high) values of effluent concentration.

Several investigators (see Orlob, 1959 and Wiegel, 1964 for summaries) have experimentally investigated the horizontal coefficient of eddy diffusion and suggested

$$D_y = eL^n$$

where L is some measure of the mean scale of turbulent eddies in which the diffusion occurs, and e and n are experimentally determined constants. Orlob (1959) found that initially $n=4/3$ but at large distances from the effluent source, D_y was effectively constant. From his and other experimental data,

Orlob recommended

$$D_y = 0.01 L^{4/3} \tag{8.11}$$

where L is in cm and D_y in cm^2/sec.

Brooks (1960) solved Eq. 8.10 with D_y given by Eq. 8.11 to obtain (with slight modification)

$$\frac{c_m}{c_0} = \text{erf}\sqrt{\frac{1.5}{\left(1 + \dfrac{8 D_y t}{L^2}\right)^3 - 1}} \tag{8.12}$$

where erf is the standard error function, c_0 is the initial effluent concentration, and c_m is the downstream centerline effluent concentration at time t.

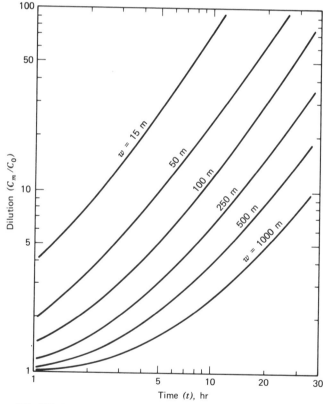

Figure 8.5 Dilution in a current for various initial plume widths (Eq. 8.12).

The usual approach for design purposes is to let L equal the projected diffuser width w. Equation 8.12 is plotted in Fig. 8.5. Note that the increase in effluent dilution, owing to an increase in diffuser length or outfall length (thus travel time to shore), occurs at a decreasing rate.

8.4. SUBMARINE OUTFALL DESIGN CONSIDERATIONS

Of primary concern in the design of submarine outfalls is determination of the required outfall location, length, and diffuser geometry. These must be established to achieve sufficient effluent dilution and organism die-away by the time the effluent is carried by currents to any location where it would be harmful to the environment. Given sufficient information (i.e. outfall characteristics and discharge, as well as receiving water characteristics and current patterns) the dilution can be predicted from the material presented in the two previous sections.

Die-away of indicator organisms (usually some enteric bacteria such as coloforms) is usually assumed to be proportional to the instantaneous concentration of organisms, where k_c is the proportionality or die-away coefficient. This yields an exponential die-away with time, and Eq. 8.12, including both dilution and die-away, becomes

$$\frac{c_m}{c_0} = e^{-k_c t}\,\text{erf}\sqrt{\frac{1.5}{\left(1+\dfrac{8D_y t}{L^2}\right)^3 - 1}} \qquad (8.13)$$

The die-away coefficient depends on the organism type; other organisms present; water temperature, turbidity, salinity, and plankton content; and other factors. Although information on bacteria die-away is available from a number of field and laboratory studies (e.g., see Gunnerson, 1958), it is best to evaluate k_c by experiments conducted at the proposed outfall site.

Rather than making use of Eq. 8.11, it is preferable, if possible, to evaluate the coefficient of diffusion at the outfall site for a variety of current conditions. This is usually accomplished by tracing the motion and resulting lateral spread of drogues, drift cards, die, radioactive substances, or some other water-motion markers, owing to turbulent mixing. From an analogy to molecular diffusion of particles in a fluid experiencing Brownian motion, it can be shown

$$\frac{d\left(\sigma_y^2\right)}{2\,dt} \approx \frac{\Delta\left(\sigma_y^2\right)}{2\Delta t} \qquad (8.14)$$

where σ_y is the standard deviation of the lateral distribution of particles (effluent or markers) being dispersed by turbulence. The markers are released and tracked (air photographs, radar, collection on shore) and the standard deviation of their lateral spread is evaluated as a function of time to determine D_y from Eq. 8.14. For examples of specific applications see Gunnerson (1958) and Orlob (1959).

Richardson and Stommel (1948) evaluated the coefficient of diffusion by tracking the motion of a single pair of floats released from a pier. They evaluated D_y from a variation of Eq. 8.14 where

$$D_y = \frac{(l_2 - l_1)^2}{2\Delta t} \tag{8.15}$$

and l_2 and l_1 are the final and initial separations of the floats over a period of time Δt.

Much information on the local environment must be collected for use in the design, construction, and subsequent evaluation of the effectiveness of a submarine outfall. As most of this information may not be available, a monitoring program should be established. Typically, data are collected at a series of stations at and near the potential outfall site and at one central station away from the site. Preferably, four or more surveys covering the different seasons of the year would be conducted. Some of the data to be collected include:

1. *Wind velocity.* periodically (e.g. hourly) during surveys plus seasonal tabulations from some nearby location, if available.

2. *Hydrography.* depth profiles along the proposed outfall location measured during each survey.

3. *Local geology.* jetting and seismic profiles along outfall location to locate bedrock and other possible obstructions to outfall construction.

4. *Wave statistics.* seasonal tabulations (from hindcast or wave gage) and daily forecasts during construction periods.

5. *Currents.* periodic (e.g. hourly) velocity measurements at site and flow pattern measurement using drogues, drift cards, radar buoys, and so on.

6. *Die-away and diffusion.* local bacterial die-away coefficients and coefficients of eddy diffusion.

7. *Water properties.* temperature, salinity, D.O., B.O.D., turbidity, plankton content, coliform count, and others, measured periodically during each survey.

8. *Sediment properties.* benthic organisms, chemical constituents.

The hydraulic design of an outfall line and diffuser section is basically a problem in the analysis of manifold flow. An example of diffuser design calculations and related considerations is given by Rawn, Bowerman, and Brooks (1960). Besides the relatively simple straight line diffuser section, diffusers have been laid out in the form of a Y, fork, and other patterns, to increase initial effluent mixing for a range of current directions.

8.5. SUMMARY

The material presented in this chapter can be used to predict the dilution of a waste effluent discharged into a coastal or estuarine water body and transported by a current to some point of concern. In addition, it is intended that this chapter will contribute to the reader's understanding of the basic dynamics of turbulent mixing and dilution.

8.6. REFERENCES

Abraham, G. (1963), "Jet Diffusion in Stagnant Ambient Fluid," Delft Hydraulics Laboratory Publication 29, July, 183 p.

Abraham, G. (1965), "Horizontal Jets in Stagnant Fluid of Other Density," *Journal, Hydraulics Division*, American Society of Civil Engineers, July, pp. 139–154.

Brooks, N. H. (1960), "Diffusion of Sewage Effluent in an Ocean Current," *Proceedings, Conference on Waste Disposal in the Marine Environment*, Pergamon Press, New York, pp. 246–267.

Cederwall, K. (1963), "The Initial Mixing on Jet Disposal into a Recipient," Hydraulics Division Reports 14 and 15, Chalmers Institute of Technology, Goteborg, Sweden.

Fan, L. and N. H. Brooks (1969), "Numerical Solutions of Turbulent Buoyant Jet Problems," Report KH-R-18, California Institute of Technology, January, 94 p.

Gunnerson, C. G. (1958), "Sewage Disposal in Santa Monica Bay, California," *Journal, Sanitary Engineering Division*, American Society of Civil Engineers, February, paper 1534.

Hart, W. E. (1961), "Jet Discharge into a Fluid with a Density Gradient," *Journal, Hydraulics Division*, American Society of Civil Engineers, November, pp. 171–200.

Orlob, G. T. (1959), "Eddy Diffusion in Homogeneous Turbulence," *Journal, Hydraulics Division*, American Society of Civil Engineers, September, pp. 75–101.

Rawn, A. M. and H. K. Palmer (1930), "Predetermining the Extent of a Sewage Field in Sea Water," *Transactions*, American Society of Civil Engineers, Vol. 94, pp. 1036–1081.

Rawn, A. M., F. R. Bowerman and N. H. Brooks (1960), "Diffusers for Disposal of Sewage in Sea Water," *Journal, Sanitary Engineering Division*, American Society of Civil Engineers, March, pp. 65–105.

Richardson, L. F. and H. Stommel (1948), "Note on Eddy Diffusion in the Sea," *Journal of Meteorology*, October, pp. 238–240.

Sharp, J. J. (1968), "Physical Interpretation of Jet Dilution Parameters," *Journal, Sanitary Engineering Division*, American Society of Civil Engineers, February, pp. 55–64.

Wiegel, R. L. (1964), *Oceanographical Engineering*, Prentice-Hall, Englewood Cliffs, New Jersey, 532 p.

8.7. PROBLEMS

1. A fluid of density ρ is surrounded by a lighter fluid of density ρ_0. Show that the buoyant force per unit mass is given by $g\dfrac{(\rho_0-\rho)}{\rho}$.

2. A 2-m inside diameter outfall line extends offshore to a depth of 30 m. It discharges 6 m^3/sec of sewage (1.000 g/c^3) into receiving water having a density of 1.025 g/c^3. What is the dilution at the plume centerline when the effluent reaches the surface? To what depth must the outfall line be extended to achieve a 60:1 effluent dilution at the surface?

3. Another way to achieve the desired dilution in Prob. 2 would be to add a diffuser with discharge ports spaced so the individual plumes do not come together before reaching the surface. How many ports would be needed to keep the same discharge velocity as in Prob. 2 and increase the dilution to 60:1 (depth still 30 m)? The same discharge velocity would mean the same exit head loss, and thus no additional energy would be needed to maintain the flow.

4. A 1-m inside diameter outfall line extends offshore to a depth of 40 m and discharges 1 m^3/sec of sewage (1.000 g/c^3). The receiving water is stratified, with the upper layer which is 10-m thick having a density of 1.018 g/c^3, and the lower 30 m having a density of 1.025 g/c^3. Will the plume be trapped at the receiving water density discontinuity?

5. An effluent with a density of 1.042 g/c^3 is discharged at a rate of 1 m^3/sec through a 0.7-m diameter pipe with an opening located 3 m below the surface. The receiving water is a fresh water lake 60 m deep. What is the density of this effluent when it reaches the lake bottom?

6. Beginning with Fig. 8.4 show and explain all of the steps leading to the derivation of Eq. 8.8.

7. The outfall lines in Prob. 2 extend 2000 m and 3000 m from shore, respectively. If an onshore current carries the effluent to the shore at a speed of 0.2 m/sec, what are the respective effluent dilutions at the shore?

8. The ports in Prob. 3 are spaced so the buoyant jets just meet at the surface. A current moves normal to the outfall axis at just the velocity necessary to evacuate all the effluent. Determine this velocity and the effluent dilution 5000 m from the outfall.

9. Briefly explain how each of the eight types of field data (i.e. wind velocity, hydrography, etc.) listed in Section 8.4 would be used in the design and/or construction of a submarine outfall for municipal sewage.

INDEX

225